MAN, MATERIALS, AND ENVIRONMENT

A Report to the National Commission on Materials Policy

by the

STUDY COMMITTEE ON ENVIRONMENTAL ASPECTS OF A NATIONAL
MATERIALS POLICY

of the

COMMITTEE FOR INTERNATIONAL ENVIRONMENTAL PROGRAMS

Environmental Studies Board
National Academy of Sciences/National Academy of Engineering

The MIT Press Environmental Studies Series
Gordon J. F. MacDonald, consulting editor

MAN, MATERIALS, AND ENVIRONMENT

National Academy of Sciences
National Academy of Engineering

The MIT Press
Cambridge, Massachusetts,
and London, England

LIBRARY OF CONGRESS CATALOGING IN PUBLICATION DATA

Environmental Studies Board. Study Committee on Environmental
Aspects of a National Materials Policy.
Man, materials, and environment.

Bibliography: p.
1. Environmental protection--Addresses, essays, lectures.
2. Environmental policy--Addresses, essays, lectures. I. United
States. National Commission on Materials Policy. II. Title.
TD170.E58 1973   301.31   73-12715
ISBN 0-262-14019-5 (hardcover)
ISBN 0-262-64013-9 (paperback)

MIT Press

0262640139

NAS
MAN MATERIALS ENVIRON

CONTENTS

Preface x

Acknowledgments xiii

Chapter 1, Conclusions and Recommendations 1

1.1 A New Materials Policy 2
1.2 Environmental Effects 10
1.3 Remedies and Institutions 14
1.4 Summary of Detailed Recommendations: Domestic Considerations 24
1.5 International and Global Environmental Considerations 29
1.6 Research and Monitoring: Unfinished Business 34
1.7 A New Emphasis 37

Chapter 2, Study Team on Economic Implications of Environmental Quality and Materials Policy 38

2.1 Introduction 39
2.2 Policy Instruments 41
2.3 Evaluation Criteria and Tradeoffs in the Application of Policies to Certain Environmental Problems 52
2.4 Should the Overall Rate of Economic Growth Be Reduced? 61
  References 64

Chapter 3, Study Team on Environmental Problems Associated with Metallic and Nonmetallic Mineral Resources 67

3.1 Assumptions, General Conclusions, and Priorities 68
3.2 Health Effects 71
3.3 Technology Assessment and Best Available Technology 81
3.4 Land Use Planning 88
3.5 Economic Considerations 94
  References 96

Chapter 4, Study Team on Environmental Problems Associated with Fuel Materials 100

4.1 Energy Overview 101
4.2 Overall Policy Guidelines 104
4.3 Coal 107
4.4 Oil and Natural Gas 117

4.5   Uranium and Thorium   123
4.6   Hydroelectric   124
4.7   Nonfossil Sources   124
4.8   Recapitulation of Policy Criteria   126
      References   129

      Chapter 5, Study Team on Environmental Problems
      Associated with Forest Products   131

5.1   Introduction   132
5.2   Materials Policy and Environmental Policy   136
5.3   Environmental Impacts Associated with Forest
      Practices   140
5.4   Environmental Problems of Timber Processing   146
5.5   International Forestry   151
5.6   Barriers to Timber Production   154
5.7   Recommendations   157
      References   160

      Chapter 6, Study Team on Environmental Quality,
      Basic Materials Policies, and the International
      Economy   162

6.1   Introduction
6.2   Domestic Environmental Controls   163
      and Short-Term Trade Patterns   165
6.3   Long-Term Effects of Environmental
      Policies on Trade Flows   174
6.4   Special Issues on the Developing Countries   177
6.5   Special Issues on Multinational Corporations   179
6.6   Aspects of Bulk Transport and Materials Supplies   180
6.7   Recovery of Tradeable Materials   182
6.8   Transnational Pollution, International Trade,
      and Basic Materials Policy   184
      References   186

      Chapter 7, Study Team on International Legal
      Determinants of National Materials Policy   188

7.1   Introduction   189
7.2   Activities within the United States   191
7.3   Activities Abroad   193
7.4   International Environments: The Oceans   197
7.5   International Environments: The Arctic   201
      Appendix – Stockholm Conference Actions
                       Bearing on Materials Policy   202
      References   205

Chapter 8, Environmental and Resource Problems
and Policies in Japan: A Comparative Case Study   206

8.1   Introduction   207
8.2   Economic Advance and Environmental Costs   207
8.3   Basic Policy--Prime Minister Tanaka's Program
      for the Remodelling of the Japanese Islands   208
8.4   Specific Environmental Problems   209
8.5   Environmental Control   209
8.6   Efficiency and Conservation in Materials Use   210
8.7   Government Organization   210
8.8   International Resources Implications   212
8.9   Energy Strategy   212
8.10 Position in the International Raw Materials
      Situation   214
8.11 Conclusions and Recommendations   215
      Appendix--Japan's Natural Resources Situation   217
      References   225

      Index   228

PREFACE

Our nation is slowly coming to grips with the problem of how we, jointly with other nations, reconcile the pressures for economic growth and the realities of resource limitations and degradation with our quest for a dignified and satisfying existence. How do we, in the words of the National Environmental Policy Act, create and maintain conditions under which man and nature can exist in productive harmony and fulfill the social, economic, and other requirements of present and future generations of Americans? That effort is a new and urgent priority; but how can it be translated into workable practices in field, forest, mine and manufacturing plant?

This study of Man, Materials, and Environment tries to provide some answers to that question. It is not directly concerned with questions of resource depletion which have been dealt with in another study by the National Academy of Sciences.* Nor does it examine the allocation of resources on the world scene, grave as that problem may be. Rather it is concerned with the problem of enhancing our environment by taking a new look at the economic and technological processes involved in materials use by building into them safeguards designed to arrest, abate, and reverse the trend of physical and social degradation.

The study was initiated by the Committee for International Environmental Programs of the Environmental Studies Board under a contract with the National Commission on Materials Policy in May 1972. In its charge to the committee the National Commission on Materials Policy** requested that the study would

1. assess how materials policy for the United States may be affected by national environmental policies, or by international agreements;

2. evaluate the effects on United States materials policy resulting from the United Nations Conference on the Human Environment;

3. by reference to selected important materials and their flow through the environment and the economy, identify and assess the implications of alternative environmental criteria; and

---

* Elements of a National Materials Policy, A Report of the National Materials Advisory Board. NAS/NAE, 1972.

** Sec. 205, Title II, PL 91-512 states, "When used in this title, the term 'material' means natural resources intended to be utilized by industry for the production of goods, with the exclusion of food."

4. identify other issues of environmental significance that are appropriate and essential for consideration by the National Commission on Materials Policy in its report to the President and Congress, June 30, 1973.

From the outset it was apparent to the committee that the short time schedule would not permit detailed investigation of the questions and issues specifically assigned to it. However, it was believed that by marshalling the judgment of a group widely experienced in materials and environmental matters, it would be practicable to identify major issues and to suggest specific steps based upon available evidence.

Under the leadership of Dean Nathaniel Wollman of the University of New Mexico, a study committee and six working teams were assembled during the summer and early autumn of 1972. They represent a remark-able diversity of professional outlook. With the assistance of an able staff, they have put together a report which gets at the major problems and proposes positive action with respect to public policy and research.

Like other activities of the Committee for International Environ-mental Programs, the report has two distinctive characteristics. The analysis blends the competence of biological and physical scientists with those concerned with technology, human behavior, and social process. It combines in the resulting findings the perspective of national welfare with keen regard for the global environment.

The report offers the following principal conclusions:

1. It is in the national interest that policies and practices aimed at satisfying the nation's need for materials essential to social well-being should reflect and accommodate at all stages, from extraction to waste disposal, considerations of environmental cost to human health, quality of habitat, and stability of ecosystems.

2. This principle should be included in all pertinent policies and legislation stipulating the right of each citizen to a healthful environment in accordance with Principle I of the Declaration of the United Nations Conference on the Human Environment.

3. The resources of all countries should be regarded as part of an interdependent habitat rather than merely as possible sources of supply; and our national policy should therefore conform to the principles of conduct adopted by the commu-nity of nations in a common effort to protect the human habitat and its resources.

4. While we can today identify and deal with environmental problems relevant to a materials policy, and while we are prepared to propose appropriate remedies, the incomplete

state of our knowledge requires urgent, systematic expansion
of research and of the institutional arrangements needed to
widen the data base.

The Committee for International Environmental Programs endorses the
conclusions of the report and expresses its sincere appreciation to Dean
Wollman, to the members of the study committee and working teams, and to
the staff.

Gilbert F. White
Chairman, Committee for International
Environmental Programs

ACKNOWLEDGMENTS

When the chairman of a project such as this finds himself associated with an expert and active group of participants, he serves best by keeping out of the way and letting nature take its course. He possesses, however, a vantage point from which he can appreciate better than others the contributions made by various individuals.

Four groups emerge with special prominence: the members of the study committee itself, the study teams, the editorial committee, and the staff of the National Academy of Sciences.

The study committee was created by the Committee for International Environmental Programs (IEPC) of the Environmental Studies Board (ESB). IEPC Chairman Gilbert F. White not only served as acting chairman of the study committee in the first months of its life, but maintained an active interest in the committee's work until the report was completed. The membership of the study committee and study teams was assembled from universities, research institutions, industry, government, and public interest groups. The study committee included the chairmen of each study team who were responsible for the drafting of Chapters 2-8. The study committee supervised the progress of the work and was directly responsible for the drafting of Chapter 1 which synthesizes the findings in Chapters 2-7 and sets forth its own conclusions based upon, but extending the recommendations of the study teams.

Six study teams prepared Chapters 2-7; Chapter 8, presenting a case study of problems in Japan, was drafted by a one-man team, Charles S. Dennison, reporting the findings during a visit to Japan made on behalf of the study committee. Each of the study teams met for two days to discuss the substance of the chapter assigned to them. Subsequently, the chairman, usually with the assistance of a team member, prepared the chapter in several successive drafts, each draft being subjected to a critique by team members, outside readers, and finally by the study committee. Eventually, all eight chapters were reviewed by IEPC and thereafter by ESB.

I wish to express my appreciation and that of the members of the study committee to the members of IEPC and ESB for their constructive and helpful advice and suggestions, especially to ESB chairman, Gordon J. F. MacDonald and ESB members John A. Swartout and David M. Gates, and to IEPC chairman, Gilbert F. White and to IEPC members who were responsible for the final review and provided detailed comments; namely, Ivan Bennett, Francois Frenkiel, John P. Holdren, Hans H. Landsberg, Thomas F. Malone, Roger Revelle, and Frederick E. Smith. My special thanks go to the members of the study committee, particularly to the study team chairmen and those who served as their alternates: Orris C. Herfindahl* and Steve H. Hanke (Economics), James S. Bethel and Henry J. Vaux (Forestry), Arnold J. Silverman (Minerals), C. Wayne Cook (Fuels), Abram Chayes and James MacDonald (International Policy), and Ralph C.d'Arge and Ingo Walter (International Economics). Paul Bugg (Economics), David B. Brooks (Minerals), Perry Hagenstein (Forestry), and Patricia L. Rosenfield (International Policy) served with their chairmen as drafters of the initial reports from the study teams.

All of us are indebted to the members of the editorial committee: Leonard Berry, Richard S. Davidson, and Charles S. Dennison who read, criticized, and reworked materials at all stages of the project.

The staff members of the National Academy of Sciences who participated in the project assumed a wide range of responsibilities that supported the study from the date of its initiation to its formal conclusion. Not only did the staff meet with and assist each of the study teams, but they filled a central role in the recruitment of team and committee members; consultation with experts in industry, government, and academic circles; editing; and logistics. I wish to express my gratitude to Ruben S. Brown, principal project officer on whose shoulders the whole business rested, to his assistant, Genevieve Atwood, to Ralph A. Llewellyn for serving as staff officer to the fuels team and to Kay K. Ryland, research assistant to all study teams and the study committee. I want to convey my thanks for the assistance, guidance and contributions to the report made by Henry J. Kellermann, executive secretary of the Committee for International Environmental Programs, and for the suggestions, encouragement, and assistance given to us by Richard A, Carpenter, executive director of the Environmental Studies Board. I also want to express my appreciation to Merri E. Oxley for the most efficient performance of the secretarial work in which she was assisted by Frances B. Jefferson and Ralph P. Davis.

I, furthermore, want to thank Lynette Wilson and Carla Duran, of the University of New Mexico, for the secretarial and administrative responsibilities that they assumed during the course of the project.

I am also greatly indebted to a number of outside experts who assisted the committee with advice in recruiting members of the study teams in organizing the study and in reviewing materials prepared by participants and specific sections of the study itself. I wish to name, in particular, Eugene Rabinowitch of the Woodrow Wilson International Center for Scholars, who served as consultant to the study and editorial committees; to Joseph Zablotney for the paper he prepared on the transportation of oil; and to Clark Blake, U.S. Geological Survey; Blair T. Bower, Resources for the Future; Preston E. Cloud, University of California, Santa Barbara; Adrian M. Gilbert, U.S. Forest Service; Robert A. Hiatt, U.S. Department of State; Hunt Janin, Council on Environmental Quality; Hardy Shirley, President's Advisory Panel on Timber and the Environment; George H. Siehl, Library of Congress; Lee Talbot, Council on Environmental Quality; Peter Twight, National Parks

---

*We were saddened by the death of Orris C. Herfindahl who chaired the team on economic implications. Steve Hanke who had become cochairman of the economics study team when Dr. Herfindahl left the country in December, 1972 assumed the position of chairman, taking over responsibility on short notice and discharging the chairman's functions with no loss of momentum.

and Conservation Association; and Ira Wolf, U. S. Department of
State.

During the conduct of the study the committee enjoyed the counsel
of J. Kenneth Klitz, Assistant Director for the Environment, National
Commission on Materials Policy, who observed our meetings, kept us
informed of the concerns of the commission, and reminded us of the
main focus of the study whenever signs of wandering became visible.

The study committee was helped by a number of federal officials
who served as liaison with their respective agencies.

In spite of the shameless way in which the study committee invited
help and exploited what was offered, it alone is responsible for the
study--its findings, conclusions, and recommendations. We hope that
the National Commission on Materials Policy will find it useful. All
of us enjoyed the work of putting it together.

Finally, I want to thank all of our spouses for putting up with
the dislocation in routine that we all suffered and for the tolerance
with which they accepted our resentment at any distraction from completing
this report.

To all of them and to all the others who freely gave of their time
and counsel, I say thank you.

Nathaniel Wollman
Albuquerque, March 1973

STUDY COMMITTEE ON ENVIRONMENTAL ASPECTS OF A NATIONAL MATERIALS POLICY

*NATHANIEL WOLLMAN, College of Arts and Sciences, University of New
    Mexico, Albuquerque, Chairman
*LEONARD BERRY, Department of Geography, Clark University, Worcester,
    Massachusetts
 JAMES S. BETHEL, College of Forest Resources, University of Washington,
    Seattle
 REID A. BRYSON, Institute of Environmental Studies, University of
    Wisconsin, Madison
 ABRAM CHAYES, Law School of Harvard University, Cambridge, Massachusetts
 C. WAYNE COOK, Department of Range Science, Colorado State University,
    Fort Collins
*RICHARD S. DAVIDSON, Bioenvironmental Programs, Battelle Laboratories,
    Columbus, Ohio
*CHARLES S. DENNISON, New York, New York
 STEVE H. HANKE, Department of Geography and Environmental Engineering,
    Johns Hopkins University, Baltimore, Maryland
 ARTHUR D. HASLER, Laboratory of Limnology, University of Wisconsin,
    Madison
**ORRIS C. HERFINDAHL, Resources for the Future, Washington, D.C.
 THOMAS F. MALONE, University of Connecticut, Storrs
 ARNOLD J. SILVERMAN, Department of Geology, University of Montana,
    Missoula
 BRIAN J. SKINNER, Department of Geology, Yale University, New Haven,
    Connecticut

Study Committee Liaison Members

JAMES R. BALSLEY, Jr., U.S. Geological Survey, Washington, D.C.
JOSEPH W. BERG, Jr., Division of Earth Sciences, National Academy
    of Sciences
JAMES FOWLER, U.S. Agency for International Development, Washington, D.C.
JOHN G. HURLEY, Office of the Foreign Secretary, National Academy of
    Sciences
REICHEL D. KURODO, Environmental Protection Agency, Washington, D.C.
JOSEPH R. LANE, National Materials Advisory Board, National Academy
    of Sciences
HUGH H. MILLER, Office of the Foreign Secretary, National Academy of
    Sciences
NATHAN E. PROMISEL, National Materials Advisory Board, National Academy
    of Sciences
HERBERT QUINN, Jr., Environmental Protection Agency, Washington, D.C.
JAMES E. TRAYERS, Council on Environmental Quality, Washington, D.C.
ROBERT A. WEEKS, U.S. Geological Survey, Washington, D.C.

*Member of Editorial Committee
**Deceased

For the Committee for International Environmental Programs:

RUBEN S. BROWN, Principal Staff Officer
RALPH A. LLEWELLYN, Staff Officer
GENEVIEVE ATWOOD, Staff Officer
KAY K. RYLAND, Research Assistant
MERRI E. OXLEY, Project Secretary

CHAPTER 1

CONCLUSIONS AND RECOMMENDATIONS

The mandate of the National Commission on Materials Policy (under Title
II of the Resource Recovery Act of 1970) is "to enhance environmental
quality and conserve materials by developing a national materials policy
to utilize present resources and technology more efficiently, to anticipate the future materials requirements of the nation and the world, and
to make recommendations on the supply, use, recovery and disposal of materials."

A national materials policy must start with recognition that materials,
energy, land and water, population, environmental degradation, and pollution
constitute a web of intersecting elements, none of which can be viewed in
isolation. Moreover, the web ignores national boundaries as materials move
over the land, through the atmosphere, and in the waters of land and sea.
A national materials policy is, therefore, a single element of the larger
task of conserving the earth for the sustenance of man's physical, mental,
emotional, and social well-being.

A national materials policy must furthermore take into account that
governments everywhere begin to accept the concept of "Only One Earth"
as a new determinant in the making of national policy. The United Nations
Conference on the Human Environment held at Stockholm in 1972 declared
that

> A point has been reached in history when we must shape
> our actions throughout the world with a more prudent care
> for their environmental consequences.

The conference charged the nations assembled there: to safeguard the natural
resources of the earth and, to that end, to maintain, restore or improve
the capacity of the earth to produce renewable materials, and in employing
the nonrenewable resources of the earth to guard against the danger of
future exhaustion. It is within this complex of changing international
attitudes and values that domestic materials policy should be fashioned.

The study committee believes that the charge to the National Commission
on Materials Policy can and must be met. The committee believes that in
meeting this charge the United States will be confronted by decisions of
utmost gravity, decisions that certain other countries must face as well.

If we extrapolate over the next thirty or forty years, the view commonly
held in the United States that two cars in the garage mean a better level
of living than one will increasingly collide with our interest in protecting
the health and well-being of our fellow citizens.

Given the present level of technology and what may reasonably be
expected to evolve over the next decades, and given the prevailing view
that materials consumption is the way to a better life, the facts indicate
(1) materials throughput will double, and then double again, over the
next thirty or forty years, (2) the quality of ores and other natural resources will decline and readily available sources be exhausted, (3) only
by increased use of energy per unit of output and per capita will the intensity of materials throughput be maintained, and (4) the environmental
stress per unit of production will increase correspondingly.

Given the growth in population, the growth in per capita product, and the growth in environmental stress per unit of product implied by sustained movement on our present track, the environmental ills presaged for the United States cannot be completely avoided by available technology.

The study committee believes that the threat to environmental quality and resource availability, caused and compounded by our treatment and use of materials, poses a real problem and a vital national issue which calls urgently for an open-minded reexamination of certain commonly held beliefs. These beliefs are : (1) that natural resources can be used in whatever amount is evoked by public demand for goods and services as stimulated and guided by producers' efforts to enlarge their markets; (2) that improved well-being of society is adequately measured by aggregate volume of the production of goods, increased per capita use of goods, and aggregate consumption of materials and energy; and (3) that technological development should and will continue to contribute to and accelerate the increased throughput of materials per person as it has in the past.

The committee recognizes that dissents to these commonly held beliefs can be found, but they constitute a relatively small voice within the prevailing views of consumers, business, labor, and governments. First corrective steps have been taken by the Congress and other legislative bodies, but without full recognition of the profound change in values that is called for: a clear assertion of each person's right to an environment that is not only healthful but possesses a beauty that reflects regard for and insistent action to cherish and perserve its natural qualities.

This right must be given full status with other basic assumptions by the nation as it seeks to provide for the physical health, intellectual growth, economic and social well-being, and security of all citizens. The study committee believes that consideration of environmental costs should be anchored in all relevant national policies and laws having as their objective the conservation or development of any sector of the national economy, including renewable and nonrenewable materials. The study committee recognizes that there are a number of laws designed to bring a balanced consideration of environmental values with economic and social objectives. In particular, the National Environmental Policy Act of 1969, in Sec. 101, addresses the crucial conflict which we have identified: material abundance vs. environmental quality. In addition, major pollutant control laws such as the Clean Air Act Amendments of 1970, the Federal Water Pollution Control Act Amendments of 1972, and the Noise Control Act of 1972 all include strong provisions for citizen suits. These laws have been enacted only recently; their success in redirecting public policy is yet unknown; they must be strongly supported by all sectors of society.

After a reasonable period of time, if administration and adjudication of these laws do not produce the necessary readjustment, an additional anchoring of the principles of environmental protection in the fundamental law of the land may be necessary. To this end we recommend the examination of the need for and the development of both an amendment to the National Environmental Policy Act of 1969 and to the Constitution of the United States declaring that the right of an individual citizen to a safe, healthful, productive, and aesthetically and culturally pleasing environment shall not be abridged.

The proposed amendment or its legislative counterpart is based on the assumption that the nation is capable of a measured and gradual transformation of its production machinery, away from concentration on scarce materials and on accelerated use of energy, and of adjustment to standards set by our new environmental policies. These changes in production do not imply loss of economic revenue or employment, although there may be considerable shifts within the labor market. On the contrary, the promotion of economic growth in sectors selected because of low risk of environmental disruption, will safeguard the economic substance of the nation and, in the long run, enhance the quality of life. Discrimination and restraint in the use of critical materials coupled with intensified efforts to recycle or to substitute materials in short supply, will in the long run diminish our dependence on foreign resources, reduce the volume of imports and prevent escalation of economic competition with other nations for scarce materials.

The study committee assumed that a materials policy designed to protect the environment will be accompanied by a compatible population policy. Population pressures being a factor contributing significantly to our environmental problems, a population policy would seem to be essential, if indeed not an indispensable complement of policies, such as a materials policy, dealing with other urgent environmental problems. The fact that the committee does not address itself to population thus is not an indication that the topic is of less importance but only that it falls beyond the committee's mission.

The committee reached nine additional conclusions of immediate concern to the national leadership:

1. A revised policy on materials will require the use of a wide variety of instruments and proposed innovations in government administration: taxes; environmental standards; standards of behavior and design; pricing and output regulations; licenses and permits; leasing conditions on public lands; and education and persuasion. The application of these instruments of policy, alternatively or cumulatively, should be decided solely on grounds of the merits and effectiveness of each.

2. Economic growth should be guided along a path consistent with policies designed to improve the environment. Sectors that impose minimum stress on the environment should be encouraged, those that impose severe stress should be discouraged. This guidance is achieved in part by the practices described in paragraphs #3 and #4, below, but additional steps are needed as well. Such steps will have to include a deliberate redirection, using and augmenting the market system, of the nation's productive capacity as well as a prudent, selective redirection of certain categories of demand. The net effect of such guidance will be to give

durability preference over planned obsolescence, to stimulate
use of materials and production methods that facilitate recycling, and to stimulate interest in sources of satisfaction
that reduce environmental stress. Educational and public
information programs should be used much more generously than
at present,

3.   A national materials policy must incorporate the principle
     that environmental costs, measured as the aggregate of social
     losses suffered as a consequence of impairment of the
     environment, are taken into account in the computation
     of benefits and costs of any action to extract, transport,
     process, use, or dispose of any material. In order to evaluate
     all environmental costs, especially those that are inade-
     quately reflected in market prices, a practice of full
     disclosure of the environmental effects of private as well
     as public activity should be mandatory. The approach
     taken by the National Environmental Policy Act of 1969 and
     the Technology Assessment Act of 1972 should be broadened,
     strengthened, and applied to all stages in the flow of
     materials through the economy.

4.   Although some exceptions might occasionally be justified,
     efficiency in the use of materials is most likely to be
     achieved when the costs of environmental damage are borne
     by those who are responsible for the impairment. Costs are
     to be charged to those who contribute to environmental damage
     by the levying of taxes, fines or other penalties, or by
     otherwise enforcing compliance with an environmental or
     design standard.

5.   When environmental effects are taken fully into account
     certain uses of materials will be perceived as yielding
     benefits that are trivial in comparison with their costs.
     This realization will be amplified by recognition of the
     finite dimension of a high quality environment and anti-
     cipation of the growth in its value relative to other things.

6.   Land use planning and the imaginative and discriminating
     use of a variety of devices, including appropriate incentives,
     are essential instruments for a policy designed to protect
     the quality of the atmosphere, rivers, lakes, coastal zones
     and oceans, as well as the land; and are also essential in
     the formulation of an energy policy which, in turn, is a
     major component of a national policy. Such land use planning
     must also take account of the need to relieve congestion in
     megalopoles, and to foster the spread of community development
     adapted to sound environmental standards.

7. In fulfilling the international and global obligations that go with a national materials policy the United States should take the initiative in adopting the best available practices, in stimulating the attention of others to environmental problems, in providing technological assistance to the best of its ability, and in joining with other nations in agreements to protect the air, the seas, the world's pool of genetic materials, and important landmarks and treasures of civilization that are threatened by environmental decay. In adjusting to the measures taken in this and other countries to protect the environment, the United States should not tolerate a growth in protectionist trade policies, to the detriment of its own and the world's efficient use of scarce resources.

8. As far as compatible with the national interest, the United States should embark without delay on a course that will steer clear of collision with other industrial powers bidding for environmentally attractive resources in short supply, such as low sulfur petroleum and liquid natural gas. It should seek jointly with major producers and consumers corrective multilateral solutions providing for orderly and equitable marketing arrangements and, at the same time, intensify the development of new technologies to ensure availability of needed resources.

9. The present state of knowledge about the origins and effects of environmental deterioration is so incomplete that it is impossible to assert with certainty how close or remote a crisis may be. In recognition of the complex forces that drive the ecosystems of the world, the committee urges the allocation of funds for expansion of research with all deliberate speed, immediate adoption of a broad program for the acquisition of base-line data, and creation of national and international machinery for adequate monitoring of environmental parameters, including effects on the ecosystem.

The committee recognizes that the environments of open spaces as well as those of cities are endangered. In the countryside, the effects of materials flows on human health are likely to be less than on the deterioration of natural habitat, degradation of the landscape, clouding of the atmosphere, and litter. As cities sprawl outward, linked by an increasingly complex network of interstate highways, prime agricultural land is preempted, reducing agricultural productivity in greater proportion than is indicated by the relative shift in land use. As a consequence of the shift in land use, use of fertilizer and other on-the-farm materials is increased without a corresponding increase in output. Within the cities, especially the megalopoles, the cumulative, often synergistic environmental impacts of materials use from automobiles, trucks, buses, power plants, factories, and heavy construction threaten human life, obliterate vegetation, and destroy many of the amenities of urban living. The aggregation of effects in concentrated areas provoke the environmental question asked by President Nixon in his State of the Union message of 1970:

> In the next 10 years we shall increase our wealth
> by 50 percent. The profound question is--does this
> mean we will be 50 percent richer in a real sense,
> 50 percent better off, 50 percent happier? Or does
> it mean that in the year 1980 the President stand-
> ing in this place will look back on a decade in
> which 70 percent of our people lived in metropoli-
> tan areas choked by traffic, suffocated by smog,
> poisoned by water, deafened by noise and terrorized
> by crime?

Environmental damage not only affects the physical-biological realm
but creates serious socio-economic problems as well. There is evidence
that those who are poor suffer more from environmental degradation
than those who are rich. Loss of environmental quality, therefore,
accentuates the inequality of income distribution and aggravates the
problems of poverty in the countryside as well as in the urban ghetto.

                              *   *   *

The conclusion that a materials policy can give due regard to the
environment is based upon findings taken from succeeding chapters,
supplemented by appreciation of the tradition of adaptability revealed
by our nation's history:

> While we have not yet explored the full range of adjustments
> of which we are capable, we know that our economic system possesses
> flexibility and unused capacity:
>
>   -- for many materials there are substitutes in
>   each of their uses;
>
>   -- certain major materials such as lumber, coal,
>   and petroleum, are available from various
>   sources, exploitation of which imposes a range
>   of stresses from relatively little to relatively
>   great;
>
>   -- the assortment of goods and services for con-
>   sumption can be changed by prudent cuts in, or
>   shifts from, consumption of environmentally
>   degrading goods and services to others that
>   are less damaging;
>
>   -- domestic as well as international economic
>   impacts of environmental protective policies
>   can be borne by adjustments in exchange rates,
>   fiscal policies, monetary policies, trade
>   patterns, and consumer preferences.

Present and new technologies, if properly applied and fully exploited,
will go far toward relieving present environmental stress. Institutions
and instruments of social control, such as taxes, prohibitions, licenses,
etc., are available to implement remedial actions.

These factors permit the hope that given the attention and priority urged by the committee there is time to adopt suitable practices that are now available, and to develop others not now known before further serious shortages of materials and irreversible degradation of the environment occur.

The presence of favorable factors does not diminish but enhances the urgency for timely action. Yet, unfortunately there are reasons to believe that we have underestimated the need for prompt action.

Many environmental hazards are only dimly visible at this time. There is still a great deal we do not know about the interaction among materials uses, the environment, the ecosystem, and man.

Biologically and geologically, as measured against the yardstick of natural processes, man has become a force to be reckoned with, not merely locally but regionally and in some cases globally. This already unprecedented absolute impact may be doubling as rapidly as every fifteen to twenty years, justifying cause for concern: corrective action may come too slowly to avert much graver environmental degradation than has already occurred.

At least three factors are operating to make it possible, even likely, that some environmental disruption will be experienced as sudden catastrophe rather than as a slowly moving and predictably growing shadow. One is the long time lag that often intervenes between an environmental insult and the appearance of harm--e.g., carcinogens and mutagens. The second factor is the irreversibility of some kinds of damage--e.g., the loss of genetic variability. The third is the apparent existence of environmental thresholds in some processes, wherein a small increment of some input to the environment generates a disproportionate response--e.g., triggering of climatic change by natural processes.

Many foreseeable problems cannot now be solved by available technology. Even if we control 99.5% of some pollutants, the remaining one-half of one percent, because of large absolute amounts projected by the year 2000, can create environmental problems for which a workable remedy has not yet emerged from the laboratory. An example is the difficulty of removing very small particles from flue gas.

Most people are not aware of the dangers to local, national and world well-being created by environmental

stresses. Most people, moreover, are unaware of the kinds
of choices available to them in conducting their affairs
so that environmental stress is minimized. Unless the
perception of environmental action is sharpened and
acceptance of remedial action is increased, we may suffer
unwanted irreversible environmental damage.

A materials policy that gives due regard to the environment will
manifest itself in a number of ways:

Explicit attention will be given to environmental
management at every stage of the materials throughput, thus
changing many decisions and actions from what they have been
in the absence of such attention.

A growing portion of materials and other resources will
be dedicated to improving the environment. The assignment
of resources to protect and improve the nation's great
natural and cultural inheritance will measurably improve
the quality of life for its citizens. The careful restoration
of the countryside or urban areas, the provision of adequate
housing, of educational and recreational facilities, and different
modes of transportation could alter the nation's bill of goods
in a most constructive way.

Some environmental problems, especially those dispersed over wide areas, can be solved only by curtailing
or eliminating the use of certain materials. For some
materials or commodities an evaluation of net benefits
yielded by their use is sufficiently inconclusive to suggest the need for further investigation. We know enough
about other materials, however, to recognize that we should
reduce or eliminate their use as quickly as possible.

When environmental effects are taken fully into account,
practices now taken for granted are likely to be reexamined.
There is clear evidence that certain practices related to product novelty, obsolescence and packaging, impose environmental
burdens excessive relative to the marginal benefits they yield.
The use of energy in housing, production and transportation must
be assessed against environmental and other costs. The automobile
with all of its contributions to American society must be examined
from an environmental perspective and its advantages weighed
against those of modern economical mass transit systems.

All resources of private as well as public sectors will
be mobilized to accomplish the education required to effect
the changed view of materials use that is needed.

To hold that every decision, public, or private, regarding materials usage must be made in light of its environmental impact implies, first, a common sensitivity to the quality of the environment; second, general recognition that the quality of the environment is closely related to man's well-being; and, third, common acceptance of the ethical rule that protection of the environment must constrain materials usage. As valuation of environmental quality sharpens and spreads, the desired mix of goods and services will change. Those who themselves put a high value on environmental quality believe that the taste for clean air, a sparkling stream, and an undisturbed hillside sharpens rather than dulls with exposure. Since the supply of natural amenities has been diminishing, the pressure of demand on remaining supply presumably will increase, not only because of population growth but also because of increased perception, justifying intensified policies today to protect natural assets of the future.

*   *   *   *   *

The remainder of this chapter is based on and summarizes succeeding chapters; a description of the damages attributable to material flows; remedies and institutions; major recommendations; international considerations, and necessary research.

1.2 Environmental Effects

The environmental effects of materials flows can be described by reference to nine major elements:

1. material: metals, nonmetallic minerals, energy materials, forest and agricultural products, polymers, ceramics, and fibers;

2. stage of the materials cycle: exploration, extraction (harvest), transport, processing, use, recycling, and ultimate disposal;

3. form of environmental disturbance: emission of unwanted substance, disturbance of surface or underground configuration, unwanted dispersal of materials, reduction of ecosystem productivity;

4. environmental medium that is abused: biota, atmosphere, land surface, subsurface land or water, rivers and lakes, coastal zones, oceans;

5. geographic character of the source of the disturbance: point source or dispersed source;

6. geographic character of the effect of the disturbance: migratory or _in situ_; within the transport medium, the sink, or both; local, regional, national, transnational, or global;

7. character of the damage to human welfare: direct assaults on life and health, direct damage to crops and property, sociocultural impact, aesthetics and related components of quality of life, indirect impact through interference with beneficial functions of ecosystems;

8. magnitude or severity, including cumulative and synergistic effects;

9. temporal factors: rate of growth of impact, continuous disruption of discrete incidents, degree of irreversibility.

A materials policy that gives due regard to environmental effects is indeed a complex matter when full account is taken of the number of combinations yielded by these nine categories for all materials, all stages of use, all environmental media, all sources, and all effects.

cf. 3.1.3   Some materials create environmental changes of a global dimension.
7.2   Relatively little is known of the magnitude of damage that these changes presage. The discharge of waste heat by manufacturing and power plants (especially over large urban regions), acidification of rainfall, and increased atmospheric burden of carbon dioxide are examples of global disturbances created by the combustion of fossil fuels.

cf. 3.3.3   Phosphorus, which is essential on a global scale for food production,
5.3.5   and recyclable only over extremely long periods of time, should be husbanded in the interests of direct and indirect environmental impacts. Directly, improper uses of phosphorus contribute to enrichment of nutrients in water and accelerated rate of eutrophication. Indirectly, the imprudent depletion of high concentration phosphate deposits would lead to increased deterioration as more land is mined for lower quality deposits.

cf. chapters   Some materials in their manipulation by man create offenses at
3, 4, and 5   successive points in the flow from extraction through processing to use and disposal. Other products impose an environmental burden at selected points of the materials flow. Gold and silver, for example, impose negligible burdens in their use or ultimate disposal. Coal, lead, mercury, cadmium, and radioactive minerals, however, place environmental burdens at successive stages from extraction to disposal. Timber imposes a burden at the time of harvest, when processed into paper or board, and when cut and planed. Secondary burdens result from the use, reuse and disposal of products. In some instances, the environmental burden is the result of unwanted concentration; e.g., tailings, spoil, slash, saw-dust. In other cases, the environmental burden is the result of

unwanted movement; e.g., atmospheric transport of heavy metals and sulfur, aquatic transport of fertilizers and pesticides, and the disposition of litter.

cf. 3.2
(all
sections);
4.3.5; 4.8;
6.2.1; 7.4;
8.4; 8.10.1

1.2.1 Effects on Human Health

Certain environmental effects directly damage the health of man. Mild or acute poisoning, respiratory disease, cardiovascular failure, and cancer can be caused by various materials. In evaluating the effects on health of emissions of materials to the environment in excess of natural levels two essential principles apply:

cf. 3.2.1
3.2.11

cf. 3.2.3

1. In the absence of evidence to the contrary, for a population of various ages and initial states of health, no threshold should be stipulated below which exposure is harmless. Instead, the response to exposure should be assumed to be directly related to successively greater or lesser concentrations of the toxic materials and the level of resistance of those exposed. Those who are most susceptible are affected by concentrations that in the short term appear to be harmless to others. Measurements of mortality or morbidity from exposure to a particular substance may conceal the fact that the measurements are made on a hardy segment of the population.

cf. 3.2.2
3.2.7
3.2.11

2. The concept of total body burden should be the significant indicator of exposure, rather than burden acquired in one or another part of the environment or from one or another toxic material. People who work in a factory in which dangerous substances are handled in high concentration, may live in an adjacent area in which the same or other substances are dispersed, thus increasing overall exposure. More than one organ may be attacked because the offending substance is transported by two or more media. Synergistic effects among two or more substances, by which the combined effect is more than the sum of the separate effects, should be considered. It is recognized, of course, that standards of health and safety may be based on the determination that protective devices may fall short of perfection. Incremental costs may prove too high if compared to the increment in protection that can be achieved.

cf. 3.2.8-
11;
4.2.3;
5.3(all
sections)

Impairments of the environment may affect man directly through health or aesthetic satisfaction, or indirectly by reduction in the vitality of vegetation and animal life and consequent impairment of the services provided to mankind by natural processes (maintenance of soil fertility, natural pest and disease control, waste disposal, fish and wildlife

production and others). The committee finds that knowledge of the
effects of exposure to various emissions is incomplete, as is also the
knowledge of other environmental disturbances that move slowly through
cf. 2.4.1   the physical-biological world. We have relatively little knowledge
   3.2.4   of the health effects of many substances, especially if contacts are
   3.2.11   at a low but sustained level and the substance acts slowly.

1.2.2 Other Environmental Effects

Scientific knowledge of the behavior of temperate forests has grown
cf. 5.3   substantially in recent years but not rapidly enough to eliminate debate
(all   regarding the effects of various forestry practices on nutrient cycling,
sections)   stability of soils, stream runoff, the health of fauna and flora, and
capacity for sustained productivity. Knowledge of the behavior of
tropical forests is even more deficient, a lack that is especially
threatening in light of prospective increase in the exploitation of
cf.   tropical forests. Possible changes in climate induced by emissions
  5.5 -   of heat, gases, and particulate materials have already been mentioned.
  5.5.4   We may also face the threat of substantial, irreversible changes in
the productivity of the coastal zones and oceans, resulting from
cf. 5.3;   accelerated deposition of waste materials and excessive harvest of
6.1; 7.4   aquatic life.

Appreciation of environmental effects of materials uses is made
difficult by interactions consequent to most decisions. Sulfur
emissions can be reduced by making greater use of western domestic
coals, but only at the expense of stripping away the land cover in an
arid region where recovery is especially slow and requiring more transportation of fuel or extension of power corridors. Motor vehicle
emissions of lead from internal combustion engines can be reduced or
eliminated but at the cost of using more fuel. Hydropower can be
produced, but only at the cost of inundated valleys and restriction
of free-flowing streams. The production of pollution control equipment is itself a strain on the environment, albeit net effects are
beneficial. Relationships among materials uses and environmental
effects become more complex as the volume of mobile pollutants increases, thereby increasing the likelihood of compound and synergistic
effects in any given place. Moreover, pollutants move from one medium
to another, such as the transport of mercury by air from the stack of
a power plant to a nearby lake, where it is deposited and converted
to toxic compounds by microorganisms and concentrated in other
organisms.

The movement of unwanted materials from one medium to another,
concentration of poisons in sinks or by animals in the food-chain, and
chemical-physical-biological changes in soils and surface waters are
forms of environmental change that affect the "productivity" of
nature as viewed by man. In some cases we measure "productivity" in

terms of a particular cultivable "crop," whether faunal or floral. In
other cases we consider productivity in terms of the health of an ecosystem with its diversity of plants and animals, such as in a forest
or lake, or the aesthetic pleasure that may be derived from such abun-
cf. chapter   dance. Environmental impacts that affect productivity and those that
    5   create aesthetic responses are likely to be intertwined, such as the
effects of herbicides, whether evaluated in terms of the target area
or misdirected use, or the effects of clearcutting and disposition of
slash on a steep mountainside.

cf. 2.3.5;   The scale of environmental impact of materials uses coupled with
3.3.1;   modern technology--whether a strip mine, a cleared forest, a complex
4.1; 8.3   of power plants, or an oil spill; or of disturbance in an environment
that hitherto had been hostile to the sustained presence of man,
namely polar region and desert; or by emission of large volumes of
substances into global media; or of complexity and persistence, such
as emissions of radioactive materials and materials that interact
adversely rather than benignly--is vastly different from anything that
has gone before. No part of the world remains unthreatened. All parts
are threatened by large scale effects. The fact that we have inadequate
knowledge of what happened in the past, is now happening, and will
happen in the future argues for caution. Evidence of the possibility
of global disaster may be inconclusive, but evidence of scarred landscapes, air pollution, water pollution, litter, depleted stocks of
fish and wildlife, environmentally provoked morbidity and mortality,
and fewer natural refuges to which man may repair are all directly
attributable to the ways in which we extract and use minerals and forest
products. The study committee believes that present concern for the
environment is neither misplaced nor exaggerated.

1.3 Remedies and Institutions

The remedies for preventing or repairing environmental damage take one
of several forms: (1) reduction of emissions; (2) dispersal of
emissions; (3) improved protection of the landscape and ecosystem;
(4) repair of the landscape and ecosystem that has been damaged by
past neglect; (5) optimum siting of unavoidable damage; (6) concentration and safe disposal. Each of these remedies may be achieved in
several ways. For example, improved protection of the landscape and
ecosystem can be achieved by stipulating standards of exploration,
extraction or harvest, zoning of activities, requirements or incentives
for rehabilitation of sites, or direct governmental investment in
rehabilitation.

Depending upon the form of the remedy and the antecedent form of
environmental impairment, the remedy is attainable by one or more types
of institutional control. For example, the threat of private suit
for damages can inhibit environmentally damaging behavior. More
commonly, the devices used are likely to be legal prohibitions of

undesired activity, compliance with affirmative standards as a condition
of undertaking certain activities, taxes, subsidies, and direct governmental action. The choice of control technique is guided by the principles that apply to the choice of solutions of social problems generally;
i.e., efficiency in use of resources to achieve established social ends,
economy and simplicity of administration, and conformance with recognized
principles of equity.

The remainder of this section is devoted to a discussion of selected
remedies and institutional devices, and to special consideration of
environmental problems posed by minerals, fuels, and forest resources.

cf. 2.2.9;   1.3.1 Emissions Taxes
   2.3.5;
  6.1;6.4   The harmful effects of point-source effluents, whether gaseous, liquid,
or solid, and whether omitted to the atmosphere, bodies of water, or
directly to the land, may be reduced by treatment at the "end of the
pipe," by changes in processing at one or more points prior to final
emission of the waste, by relocation to another point where the harmful
effects may be less, by changes in product mix, or by reduction or
cessation of operations. Common instances of unwanted emissions related
to materials are sulfur dioxide, various particulates, heavy metals,
municipal and industrial organic wastes, and heat. These emissions
accompany a wide spectrum of milling, smelting, processing, and manufacturing activities and originate in the combustion of fossil fuels,
use of nuclear fuel, and in the successive steps by which minerals,
timber, and fibers are converted into finished products.

Because point-source emissions are identified with particular
emitters and because there are various stages in the production process
where reductions can be achieved, the study committee believes that the
most efficient instrument of control is an effluent or emissions tax.
In the case of a paper mill, for example, the imposition of a tax
leaves to the discretion of the emitter at what stage of production
the objectionable material is either modified, contained, or discharged
and taxed. Moreover, the tax does not require that all emitters
behave in the same fashion regardless of the marginal costs of abatement. Finally, the tax leaves to the emitter the choices of abatement,
relocation, cessation of production, or payment of the tax. The search
for new technology is uninhibited as to locus of impact, and the desire
of consumers to shift from pollution-creating to pollution-avoiding
goods and services can operate freely through the effect of the tax or
costs of abatement on prices. The net effect, when all adjustments
have been completed, is a system of prices that reflects the social
costs of pollution and the balancing of these costs against the benefits
of continuing to use the pollution-creating goods and services.

The use of an effluent tax implies that the tax rate is set to induce the amount of abatement needed to achieve a selected ambient quality. The procedure for finding the proper tax level involves trial and error, based upon available knowledge of the industry's cost structure, competitive conditions, and demand for its products. By moving the tax rate up or down the desired ambient level is achieved, granting the fact that there will be some lag between action taken to change the level of tax and response by emitters.

cf. 3.1.2;   The abatement of polluting discharges per unit of product to a
    3.5;   small fraction of present emissions is, to the best of the committee's
    6.2.3   knowledge, well within the economic capacity of the country and, if
               induced by an emissions tax, involves minimal disturbance of management
               decisions. The committee recommends that a national materials policy
               should prominently incorporate an obligation to reduce or eliminate
               undesirable emission, and that the primary instrument to accomplish
               this should be an emission tax, supplemented by other instruments as
               needed.

cf. 2.2.6;   ## 1.3.2 Effluent Standards
    2.2.9;
    3.2.3;   Emissions that are determined to be dangerous to the health of humans
    3.5;   or to organisms critical for the health and well-being of man should be
    4.2.3;   controlled by flat prohibitions or a designated maximum for any emitter.
    5.2.1;   The choice of instrument will depend upon the recognized toxicity of
    6.1; 6.6   materials and the circumstances of emission. The emissions tax may not
               be suitable for elements that are serious threats to health because of
               the possible lag in time between imposition of the tax and final
               response.

               An effluent standard might also be justified if conflict among
               political jurisdictions makes a tax politically infeasible. If an
               effluent tax is eschewed by the federal government, competition among
               the states to serve as pollution havens would be attenuated by a federal
               effluent standard. However, imposition of a minimum federal effluent
               tax also serves to eliminate undesirable competition among states.

               In general, an effluent tax will do whatever an effluent standard
               will do except that aggregate social costs will be less with the tax
               than with the standard, or at worst, no higher.

cf. 2.3.1;   ## 1.3.3 Dispersion of Wastes: Remedy and Problem
3.4 (all
sections);   Dispersal of point-source emissions is a solution so long as the capacity
3.5; 5.2.1;   of the environment to absorb the stress is not exceeded. Dispersal is
5.4 (all   achieved mainly by a land use policy and the use of such instruments as
sections)   zoning, conditional leasing, geographic variations in emissions taxes,
6.3; 7.4;   permits to emit, and public investment in waste treatment and control
    8.3

facilities that offer locational advantages to those who are served. For materials that are discharged to the atmosphere, air zoning, although more difficult to administer, would be more effective. An awareness of the need for dispersal is most critical in areas subject to the compounding of effects from different kinds of emissions. Dispersal policies imply that decisions regarding land use are integrated with decisions regarding emission controls.

Unintentional dispersions of wastes pose their own problems, notably litter, fertilizers, pesticides, and mine drainage. Disturbances to the ecosystem by nondegradable (or slowly degradable) pesticides has led to their banishment in many countries. The problems of unwanted movement and dispersion comprise one of the most acutely dangerous aspects of materials use that we face. It is likely that the most feasible remedies will be redesign of the products so that they become harmless within a short time of application, or restriction or prohibition of use. Solution of the problems posed by migratory materials should receive high priority for attention within a materials policy.

cf. 3.4(all 1.3.4 <u>Land and Water Conservation and Reclamation</u>
  sections)
  4.2.3;   Another major class of environmental impairments stemming from materials
  4.3.2;   use is the disfigurement of the surface of the earth and associated
5.3(all   damage to soils, biota, and to ground and surface water. These insults
  sections) arise mainly from mining, mineral processing, and lumbering activities:
5.7   tailings, slag piles, open pits, scarred hillsides, landslides,
             accelerated erosion and sedimentation, loss of soil nutrients, and
             contamination of surface and ground waters.

The committee believes that lumbering and mining operations can continue at a satisfactory volume provided essential practices to protect and repair the environment are followed as detailed in chapters 3, 4, and 5. The main instrument of control is the establishment and enforcement of environmental quality standards.

cf. 3.4.5-8 The repair of land by proper disposition of spoil and revegetation
4.8;   of the surface is feasible in most parts of the country. <u>Where repair</u>
5.1.2;   <u>is not possible, mining or lumbering should be prohibited.</u> In the
5.3; 5.7   case of lumber, only a small percentage of commercial timber would be
            affected by such prohibition. In the case of minerals, a larger fraction
            is affected, since we still cannot adequately repair damage in arid and
            semi-arid, polar and subpolar environments. However, there has been
            enough progress in the development of suitable species of plants and
            methods to warrant a prohibition of the exploitation of nonreclaimable
            areas until these become reparable by new technology.

cf. 2.2.3;   As a condition of exploitation undertaken on public land, a perform-
2.2.11;   ance bond and a preoperation design should be required by appropriate
3.4.1-2;   authority. The same requirement should be met for exploitation on
3.4.7-8:   private land whenever a critical boundary (cf. 5.2.1) is encountered--
4.3.2;   i.e., movement of adverse effect across a property or civil juris-
4.3.6;   dictional line or to another medium. The responsible government
5.2.1;   agency would approve the plan and monitor its performance. This pro-
8.3   cedure is recommended as an addition to the environmental impact
             statement now required as a precedent to action on public lands.

             An environmental standard, is, in general, more important than
      specification of practices, since changes in technology will render
      particular practices obsolete. This conclusion should also hold for
      health and safety practices in mining and processing or manufacturing.
      Ambient standards as applied to water would prohibit, for example, the
      abandonment of mines and subsequent drainage of acid mine wastes.

             As a part of contemporary materials policy, the study committee
      recommends that steps be taken to repair damages committed in the
      past. The cost of such repair will in most instances be borne by
      government, since the legal liability for past actions of private
      enterprises now in existence is likely to be negligible. No extension
      of exploitation rights by lease or other privilege should be granted
      without enforcement of all reclamation obligations that are still in
      force as well as those newly incurred.

cf. 2.3.2;   ### 1.3.5 Land Use and Planning
3.4 (all
sections);   Closely related to the planned dispersion of emissions and to land
4.3.2;   and water reclamation is the need for national, regional, state, and
4.8;   local land use planning. A materials policy that side-steps the estab-
5.6 (all   lishment of national land use planning will not protect the environ-
sections);   ment under conditions of stress and will contribute to uncertainty in
6.3; 7.2;   the development of resources. Land use planning will identify and
8.3   protect wilderness and urban areas as well as designate appropriate
             sites for mining and harvesting timber. National planning should allow
             for regional and local upgrading either at the moment or over time as
             regionally or locally determined, subject to over-riding considerations
             of national welfare.

             Planning of land use undoubtedly will begin and may remain limited
      to a small number of relatively coarse decisions. In addition to the
      determinations made for mining, timbering, agricultural and green belt
      uses of land, consideration will have to be given to urbanization,
      transportation and utility corridors and to ways in which land use
      affects air and water quality. Displacement of agriculture by growth
      of cities and transportation and utility corridors has already been
      noted as a problem with important environmental implications related to

materials use. Conversely, such land use planning must take account of the need to relieve congestion in megalopoles and foster the spread of communities in accordance with high environmental and conservation standards. Since many decisions are already being made at federal, state, and local levels under outmoded jurisdictional and institutional arrangements to fix or limit land use, the question is not whether we undertake such assessment but whether the scope of planning is sufficiently broad and innovative to give due consideration to the labyrinth of consequences and alternatives of land use planning and the "means" by which the planning process itself can be organized and maintained effectively. We now see that piecemeal solutions frequently do not solve the problems to which they are addressed but merely move the problem from place to place.

An especially vexing conflict of land use is that between dedication to wilderness and exposure to exploitation, since economic measures-- i.e., ascertainable and measurable money costs and benefits--do not exist for all competing uses. We can develop partial money measures, such as the present market value of extractable timber over a designated period of harvest net of costs of extraction. We can also estimate the effect on costs of timber of marginal changes in supply. We cannot, however, by methods presently available, compute with comparable definitiveness the present value of a wilderness or of a change in the size of a wilderness. A governmental agency that has jurisdiction for land not only has the problem of estimating the intensity of demand for wilderness in the absence of a market but also the obligation of employing an interest rate that properly reflects the social rate of time preference. This rate may be different from prevailing market rates of discount used in business decisions. It is, of course, a fact that the total supply of virgin forest is incapable of expansion, yet an increasing number of people demand recreational services from the forest. The result is a prospective change in supply-demand relationship different from the prospective change in supply-demand relationships for timber per se, since timber can be grown more efficiently from much of the land presently harvested. Exploitation of forest land for minerals presents a foreseeable supply-demand relationship similar to that for recreation. The most important question in resolving such conflict of use is the relative availability of substitutes-- either other materials or other forms of satisfaction or other sources of each and the respective intensities of demand, taking alternatives into account. The study committee believes that the best allocation of land between wilderness and exploitative uses lies in the direction of the suggestion made in section 2.3.2. According to that suggestion the total area subject to conflict would be divided in conformance with the perceived relative intensities of demand for virgin land as wilderness compared with the demand for virgin land as a source of materials, taking alternative sources of supply for both competing demands into account. This procedure would take the place of case-by-case decisions, in which each side in each case sees itself the victim of the cumulative effect of a sequence of all-or-none decisions.

Land use planning would give due consideration to fragility of the environment; cost of repair or regeneration; potential for irreversibility; and alternative uses, relative productivity levels, and environmental
cf. 2-2.2;   impacts of each use. For particular areas a sequence of uses might be
3.4.5-6   readied, making the most of subsurface and surface resources before allowing investment in buildings and utilities that would be threatened by subsidence, flooding, or similar hazard prior to final reclamation.

A major problem of land use planning related to materials use arises from the siting of power plants. The attraction of western coal in the context of growing power needs on the Pacific Coast has threatened large parts of the Southwest with serious air pollution and disturbance of land surfaces. If all social costs were borne by the polluter, the consumers of power in southern California would be expected either to pay for a reduction in the output of pollutants or to compensate the nation for the loss of a major asset--the bright light and blue skies of the Southwest. No energy policy should be proposed that does not take these environmental costs into account.

cf. 3.4.5-   A more general problem is the failure at present to control the
6   siting of industrial plants in order to avoid the compounding or synergistic effects of various emissions. Knowledge of these effects is still limited, but enough is known to call for caution in what is allowed. Ambient qualities must be stipulated in more sophisticated ways and the rigor of controls and enforcement must be correspondingly increased.

cf.   Land use planning could employ various economic instruments to
2.3.1-3;   bring about a desirable dispersal of activity: variation in emission
3.4(all   charges according to location, sale of emission permits to the highest
  sections) bidder, conditional leasing of public lands, and zoning regulations.
6.2; 7.4   Land use planning would also protect public and private land from damages of multiple exploration for minerals. Leases for exploration could be issued by lottery or competitive bidding with a requirement that, in the case of public lands at least, results of the exploration be placed on file and made available to the public after a reasonable but relatively brief interval.

cf. 2.2.7;   ### 1.3.6 Design Standards
  2.2.8;
  2.3.5;   Certain problems of materials use cannot be satisfactorily handled by
  3.3.3;   emission taxes or effluent standards because of the nature of the source
  3.3.6;   or the difficulty of measuring the environmental damage. Under these
  6.2   circumstances a short-term remedy or "technological fix" can be employed; e.g., by using design standard for pollution abatement equipment. An example of such a standard currently employed is the emission control equipment for automobiles. Specified practices for the repair of mined or cut-over areas may also be viewed as design standards.

The utilization of design standards can be and is being extended
into other activities. A major source of materials use that contributes
to environmental degradation are containers--bottles, cans, bags, and
boxes--made of glass, steel, aluminum, plastic, wood, and paper; return-
able and throwaway; combustible and noncombustible; degradable and nondegradable. The container industry is closely related to product
identification and the monopoly of brandname, trade-mark, and advertis-
ing impact. Throughput environmental costs beginning with the raw
material and ending with the final disposition of dispersed litter of
the packaging-advertising-identification system are high, although no
one has computed them for the economy as a whole.

If "the polluter pays" principle is followed and if the assessment
of environmental damage is accurate, the consumer who uses convenient
(but environmentally degrading) packaging would pay his own way. If these
costs are at all consequential, however, the market system would provide
the consumer with alternative forms of packaging that would not be
encumbered by the costs of environmental damage. It would take experimentation to ascertain how application of the polluter pays principle
would affect the environment in this context.

If there is strong demand for environmentally detrimental packaging,
the imposition of a uniform or absolute design standard--e.g., reusable
bottles and prohibition of throwaway cans--may be the only feasible
form of control. Control over the adverse environmental effects of
advertising--whether in the form of billboards, signs, use of energy,
packaging, and use of paper and other materials that must be produced,
delivered, and discarded--may be made difficult by the fact that ad-
vertising reduces elasticity of demand and augments power to withstand
financial penalty. In the absence of responsiveness, design standards
in the form of restrictions against environmentally degrading practices,
coupled with design standards for packaging, will have to be used.

cf. 2.3.4; 1.3.7 The Disposal of Unwanted Solid Wastes
3.3(all
sections); A materials use policy must come to grips with the problem of solid
 4.3.3;   wastes. These originate at various stages of the materials cycle and
 4.8   confront the environment with two problems: unwanted dispersal and
5.4 (all   unwanted concentration.
sections);
5.7; 6.6;   Recycling, development of uses for presently unwanted materials,
 7.4; 8.5 and replacement of materials in or on the land constitute the remedies.
Because of tax advantages and perhaps advantages in freight rates
enjoyed by original materials the picture is not as clear as it might
be. Moreover, even if total costs of recycled materials are higher than
total costs of original materials, ignoring environmental effects,
inputs that adversely affect the environment, such as energy, may be
less. The desirability of eliminating litter and reducing other damages

to the environment implies that costs of using recycled materials should be computed after adjusting for the net environmental benefits yielded by using recycled materials, the reduced costs of handling solid wastes, differences in tax liability, and differences, if any, in freight rates. Whether recycling is economically feasible after such recomputation is not yet known. A materials policy should include a search for the facts and adoption of environmental protective practices supported by the facts.

A possible solution to the problem of solid waste disposal is use of a materials tax and tax rebate mechanism. Such tax would be imposed on materials at the point of extraction, just as a severance tax is used in many states in lieu of a property tax. The tax would be "rebated" to anyone who "deposits" in an environmentally acceptable depository an article containing the materials. Costs of administering such a system should be estimated and experimental models developed. Since many durable materials migrate extensively, the taxing and funding of the rebates should be federal. The system would be extended to imported materials--whether in raw or finished form--at the same rates as for domestic.

cf. 3.3.4;
3.4.1-2;
3.4.7-8;
4.3.3;
4.7.2;
4.8

Mine wastes are usually disposable in underground or surficial voids. The surface of mined areas when regenerated by vegetation or when dedicated to other socially beneficial uses presumably no longer constitutes the source of environmental problems. At present, the landscape is disturbed without complete extraction of valuable minerals. Some time in the future, if mineral prices have risen relative to the costs of labor and energy, residuals probably will be reworked, to the continued detriment of the environment. To handle this problem solely by imposition of the costs of environmental damage on the mine operator may not be a completely satisfactory solution. There will be a prospective future price of the material that would stimulate recovery of ores from tailings of low concentration at the public rate of discount but not at the going rate of interest in the business sector. Imposition of a tax may lead to a waste of energy caused by replacement of tailings before they have been reworked, only to have them reexposed when the price of the ore rises in the future.

The process of extraction and regeneration could be accelerated by a two-level price system whereby the federal government would purchase minerals that would be processed from tailings and pay a premium price to support the added costs. If the operator of the "tailing mine" were wholly independent of the operator of the regular mine, the possibility of collusion to take improper advantage of the premium price would be reduced. The premium would represent the public's valuation of a once-and-for-all environmental disruption and repair, in comparison with present conditions where tailings might be reworked once or twice before final disposition, based upon the social rate of discount.

cf. chapter 1.3.8 <u>Fuels: An Especially Complex Problem and Its Remedies</u>
4

A materials policy directed toward fuels must take into account the
environmental stresses created by: (1) extraction; (2) exposure to
accident and disease in mines; (3) emissions of chemicals and particulates to air and water; (4) emissions of waste heat; (5) oil spills and
leaks, on land and at sea; (6) visual effects of power stations,
power lines, tailings, and spoil; (7) disturbances created by roads,
exploration activity, pipelines, and drilling; and, (8) the possibility
of radioactive exposure.

In view of the expansion in the use of fuels that is expected
unless major social and technical measures are taken to change current
trends, the committee believes that an energy policy should be developed
for the United States that will be based upon due consideration of
environmental impacts, with recognition that costs of fuels and
electric energy will be correspondingly increased.

Where technology to maintain a satisfactory quality of the environment has not been developed, the extraction, transport, and use of
fuels should be prohibited. This policy further implies that those who
use the fuel may determine for themselves the degree of environmental
degradation they will suffer in order to enjoy use of fuels, but that
degradation may not be imposed on others without consent and compensation
on the basis of full awareness of what is entailed.

The growing rates of extraction and use, compared to inferred
potential resources of petroleum and natural gas, presents a special
problem for a national materials policy. Exhaustion of the world supply
of these materials could occur within so short a period of time that
ready substitutes would not be available for critical energy and
petrochemical industry use. Consequently, a government policy directed
toward elimination of large scale, highly inefficient use of petroleum
and natural gas, for which substitute energy sources are currently
available, is of prime importance. For example, replacement of oil
and gas by coal suitably prepared for clean burning should occur as
rapidly as possible.

For similar reasons a reduction in the amount of automobile fuel
consumed through smaller auto size and more efficient energy delivery
must be pursued under government leadership. These measures are needed
not only to conserve scarce materials but also to reduce emissions
into the atmosphere and environmental damage associated with the
discovery and extraction of increasingly elusive supplies of oil and
gas. The alternative to economizing steps such as these is likely to
be much more stringent control of energy use. Moreover, it is not yet
clear that stringent controls can be avoided even if all moderate steps
are adopted when account is taken of the atmosphere's capacity to
absorb waste heat without damage to life processes.

cf. chapter 1.3.9 **Renewable Resources: The Critical Need for Remedies Designed to**
    5   **Maintain Productivity**

The effects of environmental stress are especially dramatic within a
heavily cut forest during and immediately after the harvest of timber.
There are enough uncertainties regarding forestry techniques to make
hazardous any generalization regarding particular practices. A few
generalizations seem safe, however: (1) the value of timber now being
cut from some areas is less than the costs of extraction plus the costs
of regeneration and maintenance of productivity; (2) some harvesting
practices, when improperly used, threaten to reduce the productivity
of the land because of loss of soil and soil nutrients; (3) a relatively
large proportion of the nation's total forested area, especially areas
in the form of small holdings, is administered by relatively ineffective
methods when measured against the present state of knowledge; (4) we
are deficient in the knowledge of many aspects of forest management
and the behavior of the forest biome; (5) we know even less about the
management of tropical zone forests than we know about temperate zone
forests.

In addition to the obvious need for intensified research directed
toward protection of the ecosystem and the maintenance of long run
productivity, a number of devices and policies designed to reduce
environmental degradation of forest areas can be listed: (1) fragile
forest areas should be withdrawn from the timber supply base; (2) adherence to approved environmental standards should be a condition
of the right to remove timber, whether on public or private land;
(3) federal government support of forest culture where needed to:
(a) recognize the social benefits of forests for purposes other than
timber production; and, (b) offset imperfect applications of market
processes and state and local tax policies that are not suitably
designed for an activity of long nurture and correspondingly higher
risk.

1.4 Summary of Detailed Recommendations: Domestic Considerations

A national materials policy should start with the ethical principle
that a sensitive community lives in harmony with its natural surround-
ings, with the realization that the environment is a resource of
limited size, and with recognition of the imminent possibility, if not
immediate reality, of abusing our air, water, and land beyond the
limits of tolerance. Appreciation of the importance of environmental
considerations as an integral part of national policies and laws aiming
at the conservation and development of our national resources should
be made a permanent and binding requirement by codifying "man's right
to a healthful environment" in appropriate legislation and eventually
the Constitution. Educational and information programs that command
attention and raise public awareness should be expanded and raised to
a level commensurate with the importance of the subject.

1. A national materials policy should be based upon the
    principle that calculations of benefits and costs
    associated with the extraction, transport, processing,
    use, and disposal of materials, should take full account
    of the value of common property resources and of any
    change in the value of common properties resulting from
    the impact of materials on the environment; and should
    support the principle that those responsible for impair-
    ment of the environment should bear the costs of damage
    or repair. These principles should become a commonplace
    element of property rights, legislation, and adminis-
    trative practice at all levels of government. The
    difficulty of measuring benefits and costs should not
    delay adoption of these principles but suggests the
    need for continuous observation and experimentation.

2. In establishing health and safety standards, the traditional
    concept of a threshold must be modified to reflect that
    it does not represent a safe level of exposure for
    unusually sensitive members of the population (cf. 1.2.1).
    Health and safety laws and practices should also recognize
    that total body burden rather than exposure in selected
    environments should serve as the basis for establish-
    ment of safe standards. Standards should be constantly
    reviewed in light of new information regarding hazards
    and techniques of control. New mines, processing plants,
    and factories should be obliged to adopt the best avail-
    able technology, with older installations upgraded or
    abandoned at a reasonably rapid rate. Sufficient base-
    line surveys and studies of specific populations over
    time should be undertaken and monitoring networks should
    be developed and utilized to reveal threats to the
    health of man and other parts of the ecosystem not only
    from the gross changes of which we are already aware
    but also from the slow and subtle changes that are
    suspected.

3. Federal legislation should be introduced that calls
    for and provides administrative coordination of land
    use planning, atmospheric zoning, and the zoning of
    water resources at federal, state, and local levels,
    and that coordinates with the planning of transporta-
    tion systems, public utility facilities, and utility
    corridors and with the administration of environmental
    control practices. Land uses would be determined
    on the basis of productivity, reclaimability, fragility,
    and relative intensities of demand for different
    services of the natural environment. Of special
    concern are the delineation of urban areas, agri-
    cultural lands, mineral reservations, forests open

to commercial exploitation, forests devoted to uses
other than timbering, multiple purpose areas, wilderness
and parks, and especially vulnerable airsheds, drainage
basins, land forms, soils, and ecosystems or sub-
systems. Planning should take into account the prospect
of population growth, the fixity of space, and the
probable demands and capabilities of future generations.

4. A compatible national energy policy should be formulated
in conjunction with a national materials policy. An
energy policy should not be limited to evaluation
and use of available supplies at home and abroad but
should also consider controls on demand for energy
consistent with the needs of a high quality natural
environment. Federal legislation should be introduced
that calls for coordination in development of energy and
environmental policies and administrative actions.

5. Exploration, extraction, and harvesting should be restricted
to those occurrences and areas where damage to the surface
of the earth and productivity of the ecosystem can be
repaired or averted. As the technology of protection
or reclamation advances, additional resources can be
made available for development. In relatively few
instances where marginal social benefits are sufficiently
large and marginal costs of repair are sufficiently
high, some environmental damage may be tolerated.
Legislation should be introduced directing agencies
to establish environmental standards where these do not
now exist, to require performance bonds and plans for
preservation of the environment prior to startup
of activity that threatens environmental quality, and
to monitor performance. State and local governments
should be permitted to impose environmental requirements
beyond those established by the federal government.
Insofar as the law can provide, protective requirements
should be the same for actions on private as well as
public lands. Materials processing and other industrial
activities that threaten the environment should be
subject to the same restrictions as those applicable to
exploration, extraction, and harvest, to the extent that
can be required by law.

6. Economic policies should be coordinated with policies
designed to enhance the quality of the environment.
Economic controls should be used to discourage production
of goods and services that reduce environmental quality
and encourage production of those that impose minimum
environmental stress. Efforts should be made to

stimulate longevity of commodity life, reduce stylistic
obsolescence, and encourage use of materials and designs
that facilitate recycling. Price regulations, charges
for the use of public lands, tax provisions, and
government procurement, spending, and investment
policies that now stimulate use of environmentally
damaging materials at the expense of others less damaging should be abandoned except in the face of obvious
national need.

7. A national materials policy should use those adminis-
    trative devices that achieve the objective of protecting
    and enhancing the environment at least cost to society.
    For those instances in which an efficient emissions standard
    is feasible, an emissions tax is usually able to achieve
    the same environmental effect at lower aggregate cost
    and with less interference with the responsibilities
    of management. For other forms of environmental
    disturbances the appropriate device will be a pro-
    hibition, design or performance standard, sale or
    grant of permits, or a materials tax. In general,
    subsidies should be avoided because they lead to poor
    allocation of resources. However, in the face of
    unequal distribution of income, external benefits to
    be gained or costs averted, and political opposition
    to other devices, subsidies may be the most feasible
    way of achieving a better environment. Direct govern-
    ment investment is justified where economies of scale
    and external costs or benefits are significant, where
    a common property resource is involved, or where
    appropriate behavior by the private sector cannot be
    elicited by enforcement of environmental standards.

The attached Chart 1.1 lists twenty-three federal acts now in
effect that relate to environmental concerns. An "x" indicates an inter-
section between an act and a recommendation of the committee, indicating
that the recommendations are made in a political and legislative context
already prepared to cope with environmental problems. Existing legislation, however, is neither extensive enough in scope nor powerful
enough in sanctions to induce the social response that the committee
finds necessary. A national materials policy should call for a review
of existing legislation in order to ascertain the degree to which
present laws can be made more effective and where wholly new legislation
is needed.

APPLICABILITY OF SUMMARY RECOMMENDATIONS TO MAJOR ENVIRONMENT-MATERIALS LEGISLATION

Legislation	Recommendations 1. Benefit/cost of the commons	2. Health; Best available technology	3. Land Use Planning: Coordination with Environmental Planning	4. Energy Policy: Supplies, Demands, and Environment	5. Restrictions on Environmentally Damaging Activity: Extraction & Processing	6. Coordination of Environmental and Economic Policies	7. Administrative Devices
National Environmental Policy Act of 1969 (42 USC 4321)	x	x	x	x	x	x	
Federal Water Pollution Control Act (33 USC 1151)	x	x			x		x
Marine Protection Research & Sanctuaries Act of 1972 (P.L.92-532, 86 STAT. 1052)	x						x
Clean Air Act (42 USC 1857)	x	x			x		x
Solid Waste Disposal Act (42 USC 3251)	x				x	x	
Mineral Leasing Act of 1920 (30 USC 185)			x				x
Mining & Minerals Policy Act of 1970 (30 USC 21a)		x	x	x	x	x	x
Occupational Safety & Health Act of 1970 (29 USC 651)		x					x
Coal Mine Health & Safety Act of 1969 (30 USC 801)	x	x			x		x
Multiple-Use Sustained-Yield Act of 1960 (16 USC 528)	x		x	x	x		x
Wilderness Act of 1964 (16 USC 1131)	x			x	x		x
National Park Service Act of 1916 (16 USC 1)	x			x	x		x
Food, Drug & Cosmetic Act (21 USC 301)		x					x
Federal Insecticide, Fungicide & Rodenticide Act (7 USC 135)		x					x
Federal Aid Highway Act (23 USC 139)			x				x
Federal Power Act (16 USC 800)	x		x				x
Land and Water Conservation Fund Act (16 USC 460)				x			x
Oil pollution Act of 1961 (33 USC 1001)					x		x
Endangered Species Conservation Act of 1969 (16 USC 668)	x						x
Taylor Grazing Act (43 USC 315)				x			x
Fish & Wildlife Coordination Act (16 USC 661)	x		x	x			x
Wild and Scenic Rivers Act (16 USC 1271)	x			x			
Outer Continental Shelf Land Act (43 USC 1331)			x				x

Chart 1.1

1.5   International and Global Environmental Considerations

cf. 5.5.1;   The United Nations Conference on the Human Environment, held at
  5.5.4;   Stockholm in June 1972, epitomized the evolution of the environment
  6.8; 7.3; from a local to a national and eventually to an international concern.
  8.10.2   A series of determinants converged to produce this result.

Internationalization reflects growing awareness of the detri-
mantal effects of economic and technological development upon the
quality of the environment, the adequacy of natural resources, and
the threatened extinction of species. Inadequate anticipation of
world-wide, long run supply and demand relationships, especially
in light of growing demands of developing countries, is held responsible for jeopardizing adequate supply and for premature depletion
of basic materials needed by more slowly developing nations.

cf. chapter   Governments have realized that international action to deal with
  7   environmental problems at the source and at critical points of
manifestation is needed. Such action has been initiated through
established international channels and organizations with appropriate
modifications of original mandate and machinery (e.g., Organization for
Economic Cooperation and Development, North Atlantic Treaty Organization,
Economic Commission for Europe), through members of the scientific
community (e.g., Scientific Committee on Problems of the Environment),
and through the creation of new institutions and mechanisms (e.g., the
Stockholm Conference and the new environmental bodies generated there).
The result has been a host of new conventions, agreements, resolutions,
and recommendations designed to enlist the cooperation of governments
and nongovernmental organizations in support of specific programs.
In contrast to national policy, the international initiatives are
neither legally binding nor enforceable but rest largely on the
power of moral suasion, on the determination of governments to honor
their commitments, and on the ability of governments to marshall
domestic support for their commitments. The international innovations
have created, nevertheless, a new political climate, a new set of
political ground rules, and new principles of international conduct
which governments will find increasingly difficult to ignore or defy.

Third, the establishment of national environmental controls is
bound to have a variety of transnational effects of a political, economic,
and technological nature. The effects can be positive or negative.
National controls may move other nations to agree to the adoption of
common standards, policies, and practices or may provoke counter-
action designed to neutralize and even combat the effects of controls
outside the country of origin, especially if controls are suspected
of passing on the problem to other countries (through the establishment of pollution havens) or placing the latter in a position of
competitive disadvantage. The uncertainty of reaction--whether
supportive or competitive--suggests the need for international
instrumentalities to deal with contingencies promptly and effectively.

Formal international environmental actions antedate Stockholm by nearly two decades.* Moreover, the growing injection of environmental considerations into programs originally designed for wholly different purposes demonstrates that the United Nations will have to share credit and responsibility for environmental initiatives with a host of other sponsors.

cf. chapter 7

The Stockholm Conference presented the world with the most comprehensive action program and the broadest spectrum of international sponsorship to date. As one of the most active participants and the original sponsor of the resolution which established the new executive structure, the United States has assumed special responsibility for continuing support as well as an obligation to bring about, as far as possible, a harmonization of national policies with the principles enunciated at Stockholm. The most important principles adopted in Stockholm and relevant to the requirements of a national materials policy are those concerned with the protection and rational management of the earth's natural resources and environment, control of the discharge of toxic substances and heat, protection of the seas, and the integration of ecological considerations in development planning. Other relevant principles include application of science and technology to the identification and control of environmental risks; education, research, and development with respect to environmental problems and their management; and the development of international law to compensate for environmental damage caused by states in areas outside their jurisdiction. The 109 recommendations adopted either unanimously or by majority decisions were organized in five principal categories: planning and management of human settlements for environmental quality; environmental aspects of natural resources management; identification and control of broad international significance; educational, informational, social and cultural aspects of environmental issues; and development and environment. In their aggregate they form the so-called "Action Plan". They are addressed for purposes of implementation to governments, to United Nations bodies and agencies, and to nongovernmental organizations. They are binding only upon United Nations bodies. However, a special Resolution on Institutional and Financial Arrangements provides new machinery for joint action and coordination by governments, United Nations agencies and nongovernmental organizations.

----

* In 1940 the United States acceded to the Convention on Nature Protection and Wildlife Preservation; in 1944 it concluded the International Whaling Convention, and in 1954, the International Convention for the Prevention of Pollution of the Sea by Oil.

In its relations with other countries, whether bilaterally, multilaterally, or through participation in international agencies, the committee expects the United States to follow practices that preserve its own interests as well as those of the world at large. This coincidence of interests is most likely to be attained if actions of the United States in pursuit of a national materials policy conform to generally accepted principles of international relations, political and economic, whereby the United States:

cf. 7.3   1. supports the principle that man has a fundamental right to adequate conditions of life in an environment of quality;

cf. 7.1   2. recognizes its responsibility to help safeguard the natural resources, including air, water, and land; and to help preserve the nonrenewable resources of the earth.

cf. 7.4   3. joins with other nations in efforts to prevent the pollution of the seas;

cf. 6.4; 8.9   4. supports the right of each nation to decide for itself, insofar as the decision has no adverse external environmental effects, the quality of its own environment and the steps taken in conformance with that decision;

cf. 6.4; 7.3   5. makes available full information to any country regarding the quality of and potential threats to the environment as related to any practice regarding materials extraction, processing, transportation, use, or ultimate disposition;

cf. 6.1;   6. supports general adoption of the "polluter-pays" principle
6.2.3;   as part of a materials policy, but accepts and takes no
6.4; 7.1;   retaliatory action against the products of a nation that
7.2   follows other practices in financing its own environmental
                  control programs;

cf. 6.4; 7.1; 7. supports the principle that in accordance with the Charter
7.3; 8.10.2   of the United Nations and the principles of international
                   law, nations have the responsibility: (1) to ensure
                   that activities within their jurisdiction or control do
                   not cause damage to the environment of other nations or
                   areas beyond the limits of national jurisdiction; and (2)
                   to cooperate in the development of international law regarding
                   liability and compensation for the victims of pollution
                   and other environmental damage caused by such activities;

cf. 6.1; 7.2   8. adjusts to business decline or persistent balance of payments
8.9-10   disequilibrium as a consequence of environmental controls

and increased competition from abroad by use of general fiscal and monetary devices, labor retraining and relocating programs, and other practices that do not offset comparative advantage as revealed after accounting for the environmental standards established by each country and the ease with which they can be achieved. Import restrictions should be used only in areas where required to protect the domestic environment (of the United States) and should be applied on a nondiscrimatory basis relative to materials and products of domestic origin;

cf. 6.4; 7.3   9.   follows procedures which ensure that environmental impacts of foreign aid projects are brought to the attention of recipient countries and that plans and analyses reflecting consideration of such impacts are worked out in close cooperation with those countries, thus avoiding any implication of coercive pressure on host countries or the imposition of United States standards, priorities or procedures on the decisions of foreign governments having domestic importance only;

cf. 5.5.3;   10.   participates in the development of an equitable arrangement
7.3   for the payment of compensation to habitat countries when global decisions to protect certain species impose a consequential economic burden on the country;

cf. 5.5.3;   11.   joins with other nations in mutual agreement to avoid
7.3   environmental hazards and to protect endangered species, recognizing the restrictions that such agreement may impose on the exercise of the will of sovereign powers;

cf. 7.2   12.   vests the United States-Canada International Joint Commission and the Boundary and Water Commission, United States and Mexico, with comprehensive authority for environmental matters concerning boundary waters and airsheds;

cf. 7.3   13.   encourages the preparation of environmental impact assessments by international financing agencies;

cf. 6.6; 7.4   14.   facilitates improved environmental protection of the oceans by supporting negotiations for enforceable prohibitions against threatening practices, and by supporting steps to define the boundaries of territorial and adjacent waters more precisely;

cf. 7.5   15.   joins with Canada and other appropriate and interest nations in taking whatever steps are necessary on binational and multinational bases to protect the environment of the Arctic Basin.

cf. chapter
8

Competition in world markets for materials that contribute
minimum environmental damage while meeting industrial needs will
become more intense in decades to come. Japan and the United States,
and both vis-a-vis western Europe, are likely to be the main competitors as each seeks to accommodate to its respective aspirations.
In anticipation of an ultimate equilibrium among competing forces,
the study committee recommends several initial steps to be taken
by the United States in its relationship with Japan with extension
to other countries as appropriate:

1. that current consultative procedures between both govern-
    ments be continued and extended at all levels--executive,
    foreign policy, economic policy, materials policy,
    environmental policy, and on specific technical problems;

2. that both countries give close consideration to the implications
    of growing competition between them for supplies of certain
    desirable raw materials and that both work toward an
    expanding, unhampered trade system;

3. that both countries seek to harmonize their relations
    with raw materials-producing developing countries, parti-
    cularly in matters of development and technical assistance,
    thus avoiding the tensions and risks of power rivalry;

4. that both countries use their world influence to support
    the international organization and programs proposed at the
    United Nations Conference on the Human Environment at
    Stockholm.

cf. 6.5;
7.3; 8.7;
8.9-8.10

United States multinational corporations are likely to maintain
a major role in the transmission of technology and the convergence of
environmental standards of developing and developed countries.
The committee recognizes that multinational corporations are subject
to local environmental policies and competition with corporations of
other countries. In spite of these constraints, because of size and
importance in world markets, multinational corporate enterprise is
able to exercise freedom in the degree of concern it displays toward
the environment and the way it exerts its own influence regarding
the establishment of environmental policies in countries of which
it is a resident. Private enterprise, in general, has freedom in
deciding which practices are proprietary and which, as they relate to
technology applicable to the environment, can be put into the public
domain. Multinational corporations should assume responsibility for
disclosing fully the prospective environmental impacts of any activity
they will undertake. If multinational corporations attend to the
environment to the maximum rather than minimum limit that circumstances
in each host country allow, their role as active participants in the
adoption of environmentally beneficial practices will be maximized.

The way in which multinational corporations can be induced to adopt practices that reflect a high level of concern for environmental quality when the host country doesnot impose environmental controls and when no public funding is entailed should be investigated as part of a national materials policy.

Except for reference to United States-Japanese competition in world markets, the committee has not addressed itself to the question of allocating world supplies among specific nations. Apart from the general supply-demand relationships that impending economic development throughout the world implies, there is the special question of environmental considerations. What machinery should be used to allocate resources that have special environmental significance? Should special machinery be used? If so, by what rules would it operate? What materials would be subject to such machinery? While these questions may not be immediately central to a national materials policy they are of great consequence and should be pursued by appropriate national and international officials.

1.6 Research and Monitoring: Unfinished Business

The pace with which action to protect the environment has grown, from an almost standing start hardly more than a decade ago, has not yet matched the speed of environmental degradation in spite of particular instances of improvement. In addition to the need for a wide range of policies, administrative actions, and remedial steps as described in the previous sections of this chapter and
cf. 2.2; elaborated upon in the team reports that follow, there is an urgent
2.4.1; need for basic research, research directed toward the solution of
3.2.4; specific problems, the acquisition of basic data, and a monitoring
5.5.4; network. An essential component of a materials policy is the flow
6.8; 7.2; of information by which management of the environment can be accom-
8.10 plished. Acquisition of this information implies orderly and close collaboration with international and other national information systems.

The various environmental impact statements of which the committee is aware do not always convey information that is needed on effects, remedies, or the ways by which remedial action is managed. In addition
cf. 4.3.5, to these gaps in knowledge it is possible that much is happening of an
5.2.1 ominous nature about which scientists are unaware or at best suspicious.
7.3

Given the state of uncertainty regarding the boundaries of knowledge, the only safe course of action is an investigative program of sufficient magnitude and intensity. The committee believes that an
cf. 3.2.3, investigative program that will provide longitudinal data (studies of
3.2.4 specific populations over time) on human health should be undertaken over a spectrum of localities ranging from those in which emissions

are known to be high to those suspected of being untouched. Without such studies it will not be possible to establish the toxic effects of low level but prolonged exposure. Federal financial support of environmentally related health studies should take cognizance of changes in congestion, vehicular traffic, meteorological conditions, and living styles that might be suspected of having a relationship with health. These environmental epidemiologic studies should incorporate sufficiently refined data on environmental characteristics to reveal changes over time; changes in total body burden; synergistic and antagonistic effects; and the classification of parameters according to relevance to health. Studies should include data on toxic materials carried by air, food, and water and how these media affect absorption by humans. Measurements should include environmental qualities of industrial plants, industrial areas outside the plants, and the community in general.

The need to control industrial emissions imposes a research and development obligation throughout all industrial stages, not only at cf. 2.2.2; the terminal stage of emissions. Once such efforts become a sus-
3.3.6; 3.5; tained obligation at all levels of industry, the costs of abatement
4.2.1; 6.8; are likely to fall below current estimates. The record to data
8.5 provides strong support for such optimism. Two of the most serious problems, because of health hazards that are entailed, that have received inadequate attention are the control of fine particulate and the understanding of synergistic effects among pollutants.

As noted elsewhere, the long-range effects on climate of chemical and particulate emissions and of heat discharges are not known, although there is sufficient basis of concern to warrant a global effort at resolution. Closely related to questions of climate are the phenomena of long distance transport of various emissions, their concentration by physical and biological processes, and their effects on rainfall and the oceans.

cf. 5.3; The contribution of improper land use to wind and water erosion
6.4; 7.4 and transport of sediment to the sea converts a domestic into an international environmental control problem. A materials policy should address itself to the nature of the research efforts that should be undertaken in different parts of the world, taking cogni-zance of climate, soils, forestry-agricultural rotations, and social-economic institutions. The way in which trees are grown and harvested cf. chapter is likely to become a matter of global concern because of the impact
5 of forestry practices on the biological productivity of large areas of the world. Productivity in this context covers not only the produc-tion of wood but also the entire forest-related ecosystem, the pro-duction of stream runoff, beneficial or adverse climatic effects, and control over the mass movement of land. A major program of basic research and refined monitoring of environmental factors on a global scale is urgent.

Research is equally urgent on problems of migratory materials
originating in agriculture and the loss of prime agricultural land,
with the consequent increase in on-the-farm use of chemicals and fuels
and their adverse environmental effects.

cf. 2.4;   A materials policy should support investigation into, simulation of,
2.4.1;   and experiments with large scale changes in the way of life and manner
3.1.2;   of materials use. Current living patterns in the western world are a
5.1.3;   direct outcome of the industrial revolution and its preoccupation with
5.7; 6.9;   the mass production of fuels and materials. The forms of urban dis-
8.2; 8.9   turbance that we currently experience are the product of antecedent
forces that have been at work long enough to provide a sense of
historical necessity yet short enough to indicate that they might be
only a passing phase. If present trends of materials use are extrap-
olated into the future, even if there were to be a drastic reduction
in the rate of population growth, the results of the doubling of production every fifteen or twenty years on a global scale will clearly
be disastrous. The question is, what can be done? A materials policy
should support a number of investigations and models that test radically
new solutions: urban design and transportation networks; energy demand
and production and distribution systems; packaging-marketing-advertising
systems; nonmaterials-oriented recreational activities; design changes
that increase durability, eliminate planned obsolescence, and rely
upon materials that impose minimum stress on the environment. The
committee recommends study of national economic models that are dis-
aggregated into regions and that incorporate alternative environmental
standards and population growth rates.

cf. chapter   Among the new systems that urgently call for study, that of
4   energy, is paramount. Limited evidence suggests that basic research
and experimentation with solar and geothermal energy should be substantially increased. Although utilization of these forms of energy will
pose environmental problems, their use is likely to lead to a reduction
rather than increase in environmental stress. Probably the most
impressive long run prospect is held out by controlled thermonuclear
cf. 4.7.2   reaction, as an abundant source of energy and as one that could contribute minimal damage to the environment. The prospective interval of
time until fusion becomes a feasible source of energy, however,
cf. 4.3   suggests that coal will become the dominant fuel as oil and gas
(all   supplies are depleted. Research on reducing the environmental damages
sections)   created by use of coal is urgently needed, with gasification at the
mine probably being the most promising.

The need for a major increase in research support can scarcely be
overstated. If words such as "crisis" or "catastrophe" seem to be
unscientific exaggerations of the nation's environmental state of
health, it is only because of the faith that we have in technology and
social measures to reverse trends that are now ominously under way.

Because the environment is a common property resource, and because
the private sector will undertake only those activities that relate
directly to its private accounting of costs and benefits, major
responsibility for support of environmental research must rest with
the federal government. A materials policy should provide adequate
support for the study of institutional arrangements that are needed
to facilitate environmentally desirable adjustments as well as for
technological research. It should also recognize that the nation's
welfare is inextricably bound up with the state of the global environment, and that improvements in the methods of producing and using
materials in a distant place may enhance the economy and quality of
life at home.

The study committee urges Congress and administrative agencies
to increase substantially their support of research in both pure
and applied topics and to support investigations and pilot projects
in large-scale system changes. In light of research capability now
available, the level of funding for environmentally focused research
and development could probably be doubled without unduly straining
the research capacity of the nation.

1.7 A New Emphasis

Evaluation of considerations relating to a materials policy in light
of what has happened since the report to President Truman of the
President's Materials Policy Commission in 1952, William S. Paley,
Chairman, indicates where the new emphasis should be.*

A policy for the seventies, and probably thereafter as well,
will reflect awareness of prospective changes in environmental quality
associated with alternative paths of economic growth and materials
uses. The study committee supports all steps that can be taken
within the federal government, beginning with Constitutional protection
of individual rights and extending to all action that can be taken by
the executive office of the President and the Congress, to bring
environmental factors into prominent and sustained consideration as
part of the evaluation of national priorities, and to provide assurance
that decisions within the federal establishment give due regard
to the fact that an environment of high quality is a major component
of national well-being.

---

* Paley, W. S., July 9, 1972. "U. S. Agency Needed: Constant
Review of Policy is Vital," The New York Times

CHAPTER 2

STUDY TEAM

ON

ECONOMIC IMPLICATIONS OF ENVIRONMENTAL QUALITY AND MATERIALS POLICY

*ORRIS C. HERFINDAHL, Resources for the Future, Washington, D.C., Chairman
STEVE H. HANKE, Department of Geography and Environmental Engineering,
        Johns Hopkins University, Baltimore, Maryland, Cochairman
PAUL BUGG, Department of Geography and Environmental Engineering, Johns
        Hopkins University, Baltimore, Maryland
JOHN H. CUMBERLAND, Department of Economics, and Bureau of Business and
        Economic Research, University of Maryland, College Park
ROBERT K. DAVIS, National Audubon Society and Department of Geography
        and Environmental Engineering, Johns Hopkins University, Baltimore,
        Maryland
IVARS GUTMANIS, National Planning Association, Washington, D.C.
RICHARD T. NEWCOMB, School of Mines, University of West Virginia, Morgantown
GENE WUNDERLICH, Natural Resource Economics Division, U.S. Department
        of Agriculture, Washington, D.C.
RUBEN S. BROWN, Staff Officer

*Deceased

2.1 <u>Introduction</u>

Materials policy may be thought of as including all those actions of
government that affect the extraction, production, use, and disposal
of materials. This study team's concern, however, is not with the
whole of materials policy but only with that part of it that is related
to environmental problems. Relations here may flow in either of two
directions. Measures taken to solve certain environmental problems
may affect the materials economy--by changing costs, altering relative
demands, and so on. On the other side, measures taken in the
name of materials policy may have environmental repercussions, thus
influencing environmental policy. This examination has been restricted
to some general economic aspects of environmental policy that have
important implications for a national materials policy and to a
few specific cases where the relationship between environmental
policy and materials is particularly close and where economics seems
especially able to provide insight.

However, before presenting the findings, it will be useful
to present some of the study team's underlying assumptions. They are
as follows:

1. For now and for some time into the future the fundamental
   instrument for attempting to raise the value of the social
   product is the <u>market system</u>. The objective is not to oppose
   the market system or to discard it as an institution
   but to prescribe policies that will correct its imperfections
   (Bator, 1958).

2. Services provided by the environment are among the scarce
   productive resources which must be economized (Caldwell,
   1963; Kneese and Herfindahl, 1965; President's Science
   Advisory Committee, 1965).

3. Since there is a large degree of interrelatedness within
   and between the biosphere and the technosphere, it is a
   mistake to attempt to solve any resource allocation problem
   by concentrating on narrow objectives (Kneese, Ayres, and
   d'Arge, 1970).

4. The policy goal in handling environmental disruption and
   other undesirable external effects is to devise and use
   mechanisms whereby economic units are led to act as if they
   were taking into account all of the consequences of their
   actions and not just those within their own plants or house-
   holds (Hufschmidt, 1971).

5. In many cases, the most effective policy instruments for
   achieving environmental objectives are those which simulate
   the interactions of the competitive market.

6. Reduction of environmental disruption should be carried to
    the point where the cost of further reduction is greater
    than the associated gain.

7. The choice of policy device should depend on the cost with
    which a given reduction in environmental disruption can be
    achieved. These include administrative and legal costs as
    well as other costs incurred by producers and consumers
    (Demsetz, 1964).

8. Benefit-cost analysis is performed, at least implicitly,
    in all decision making. When it is employed explicitly,
    it is important that the scope and measurement of benefits
    and costs be chosen rationally. It is crucial that benefit-
    cost analyses be part of a decision making procedure that
    explicitly includes intangible or non-market values (McKean, 1958).

9. The distinction between resource requirements, which are
    independent of price and resource demands, which are a function
    of price, should be explicitly recognized in policy planning
    dealing with resource allocation (Hanke and Boland, 1971).

These assumptions led this study team to the conclusion that,
for purposes of public policy, environmental disturbance is an economic
problem. In fact, it is difficult to discuss any particular environmental disruption without reference to the damages from the disruption
imposed on present and/or future users of the environment and the costs
of avoiding them.

2.1.1 Economic Terminology for Environmental Concepts

The salient characteristic of "externalities" in the economic process
is that they provide real benefits or impose real costs on third
parties but these benefits or costs are not directly considered in
the economic calculations of the one who generates them in the course
of his activities as a producer or consumer. The classic example of
an external economy is that of the beekeeper and his neighbor, the apple
grower. The bees, in gathering nectar, pollinate the apple blossoms.
Thus, the beekeeper by locating his hives near the orchard helps to
confer an external economy on the apple grower since, of course, the
apple grower does not pay for the pollination services of the bees.
The classic example of an external diseconomy is air pollution in
the form of smoke from a factory. The smoke may impose real costs
on the surrounding countryside and its inhabitants in terms of increased
cleaning expenses, poorer health, and damaged vegetation, but these
costs do not appear directly on the accounting books of the factory.
There are also externalities in consumption as the case of discomfort
of nonsmokers confined with smokers.

The origin of many, if not all, environmental disturbance problems

can be interpreted as arising from the fact that some natural assets
are held in common, or, more accurately, are not owned by anybody--
"common property resources." The fact that these natural assets are
not owned or treated as if they were owned means that no charge is
levied for using the important services that they provide. This
results in an overuse of scarce common property resources. In the case
of the atmosphere and water bodies, this means that economic units
act as if these resources' ability to assimilate pollutants were
indefinitely large. The fact that economic units are not required
to take into account the scarce nature of environmental assets is a
general and helpful way of describing the fundamental source of
"environmental" problems.

Particular types of externalities are related to given institutional arrangements. Institutional arrangements refer to the complex
of laws, rules, regulations, and customs within which social,
political, and economic activities operate. By changing the institutional arrangements, one can frequently force the producer of an
external diseconomy to "internalize" the costs; i.e., bear the
burden of reducing the externality, so that the cost becomes part of
his direct economic calculations. (See section 6.2.3 for a discussion
of alternative institutional arrangements.) For example, if a legis-
lature passed a law requiring mine operators to pay a tax on mine
water discharged into a stream, an economist might say that the
government had changed the institutional arrangements to force the mine
operators to internalize an external diseconomy associated with a common
property resource. It was assumed by this study team that a major
goal of a national materials policy is to change institutional arrangements to force materials users to internalize within their economic
calculations external diseconomies associated with common property
resource use.

In circumstances in which the number of people involved or other
difficulties make it prohibitively costly for the polluter and the
pollutees to negotiate a settlement that would be to their mutual
advantage, intervention by government in some form or other may bring
about an improvement in affairs by requiring or stimulating reduction
of the environmental disruption. Whatever method is used, the economic
units disrupting the environment will no longer be free to ignore p
the costs that this action imposes on others. The object is to
"internalize" these costs and thus to bring them into the range of
factors considered by the polluter in making his economic decisions.

2.2 Policy Instruments

2.2.1 Considerations in Assessing Alternative Policies

In the past, many environmental and materials policies have been
formulated without making the following factors explicit:

    1. the objective of policy;

    2. the alternatives considered;

3.   the evaluation criteria; and

4.   the tradeoffs, including the cost of negotiation, involved.

The objective might be to increase the value of social product;
the alternative policies might include those discussed in the text below;
the criteria would include economic efficiency (a balancing of incremental benefits and costs, broadly defined to include transaction costs)
and income distribution; the tradeoffs would be made in terms of real
economic benefits and costs. The use of such a methodology can clarify
many issues and choices, making inherently complex situations more
tractable.

Recommendation

We recommend that each of the above elements of policy making be
considered explicitly when considering environmental and materials
policies. Each element will be discussed below.

2.2.2   The Objective of Policy--Enhancement of Ambient Quality

At the present evolutionary state in the development of environmental
policy, the primary objective is the preservation and enhancement
of the quality of the ambient environmental media, the air, water,
land, and bio-mass. For example, the States have established local
ambient quality standards, within Federally determined guidelines
and scientific criteria (Public Laws 89-234 and 91-604; Air Pollution
Control Administration, 1969-1972; Federal Water Pollution Control
Administration, 1968 and 1972). At a given time and place, there is
a maximum permissible level of discharge or disruption consistent
with such ambient standards. Moreover, there are several policy
tools that can be used singly or collectively to meet these ambient
standards.

The goal of enhancing ambient quality can only be evaluated in
terms of the alternative uses which society can determine for environmental resources. Are the damages caused by one use so severe that
they permanently preclude all other uses? Are there several less damaging
uses which can be reconciled with each other, either sequentially, in
one environmental zone, or spatially (upstream/downstream)? What
political and economic constraints affect remedial or preventive
solutions to specific environmental problems? What constraints on
the use of materials will society be willing to accept in order to
realize a sustainable level of ambient quality? The basic question
here for the economist is what is the optimum level of pollution
consistent with maximizing social welfare.

Progress toward the reduction of damages and the enhancement of
ambient quality to reach sustainable levels is not yet fully measurable.
Measurement and monitoring systems remain in their infancy, as

emphasized in the 1972 Annual Report of the Council on Environmental
Quality, in the recommendations of the Stockholm Conference, and by
the Scientific Committee on Problems of the Environment (SCOPE) of
the International Council of Scientific Unions which found

> "... that the present machinery for environmental
> management and resource exploitation is based on
> insufficient knowledge. It is not possible at
> present to provide the information we need to
> define and understand the large-scale processes
> going on in the biosphere, partly influenced by
> man's activities. We are convinced, however,
> that better definition and understanding can be
> brought about by appropriate research and monitor-
> ing. This will not only provide criteria for
> environmental management in developed countries
> but will also enable developing countries to
> avoid environmental disamenities and to establish
> a rational system of natural resource management."
> (SCOPE, 1972.)

The preservation and enhancement of ambient quality is thus the
primary objective of current environmental policies which must be
considered in devising a national materials policy, and the measurement
of ambient quality must be an integral part of any strategy to reduce
environmental disruption.

2.2.3 The Range of Choice in Policy Instruments

If a given activity is creating an environmental problem, ideally
what we would like to do is to reduce, in the least costly way, the
damages resulting from that activity. This should proceed to the
point where the cost of further reduction of damages is just balanced
by the cost from the damages that would otherwise result. In some
simple cases, it is comparatively easy to achieve this balance by
altering property rights and responsibility for damages through private
contracts and litigation (Coase, 1960). Where there are only two or
a few parties involved, for example, those being damaged can pay the
polluter to cut his emission if this really is to their advantage,
assuming that the law does not permit them to sue for damages done.
On the other hand, if the law permits a suit for damage, the polluter
may be liable to pay those damaged.

The reliance on rules affecting market mechanisms, private contracts,
and litigation thus represents one general class of policy instruments
that continue to be available and which should be improved and used
whenever possible, and cases of this type are very frequent. But they
hardly encompass all of the serious environmental problems that confront
us. The main reason they do not is that the transaction costs (which
include the costs of finding out who is damaged and organizing the
bargaining or litigation) are simply too high.

A second general class of policy instruments addressing environmental problems is regulatory action by some level of government. However, the obstacles to effective regulation are many. As already indicated, understanding of the damages themselves may be uncertain and incomplete. In addition, the environmental disturbance may be associated with the production of commodities that society wants to consume. Moreover, every regulatory device carries with it its own costs of administration. It is easy in some cases to conceive of an otherwise satisfactory regulatory scheme whose costs of administration would be so high as to be greater than the value of the damages avoided.

A variety of possible regulatory measures might, however, be adopted in any particular case. They will have different impacts on those who are being regulated. For example, some will require greater outlays by the regulated firms or consumers to achieve a given reduction in emissions than would other measures. Thus, the administrative and/or implementation costs of the different measures will not be the same. With varied prospects for success, more than one regulatory device might be applied, either simultaneously or sequentially, to achieve results in a given situation, such as tax relief, permits, and fines. Newly applied devices, such as ambient quality standards, effluent charges, and outright prohibitions, may supersede earlier ones if, and as, public, scientific, and administrative awareness increases and more rigorous ambient quality requirements are set. From the point of view of society as a whole all of the costs and benefits associated with the proposed regulatory device must be evaluated, including the administrative and legal costs associated with the regulation.

For example, no one fully anticipated the administrative, legal, informational and research costs of the impact statement procedure required by Section 102(c) of the National Environmental Policy Act of 1969. Yet the benefits for environmental enhancement appear to have been such that pressures are building up to extend to other levels of governments (States and municipalities) and to private enterprise the principle requiring the public disclosure of anticipated or observed impacts of significant actions affecting environmental quality.

The third general class of policy instruments would be direct government investment to reduce the externality. In cases in which there are economies of scale in disposal or reclamation, direct government investment might be the best way to handle the environmental problem.

The policymaker should be aware of the existence of a continuum of policy instruments among which he may choose. From the standpoint of economics, all policy instruments may be viewed as conferring a benefit or imposing a burden on the affected parties. One result is a transfer of income or wealth between the public and private interests or among private interests. The transactions costs, costs of implementation, feasibility and effectiveness, and incidence will also tend to differ.

In summary form, the various policy instruments can be arranged along a scale from total reliance on the market system accompanied by litigation to total reliance on direct government investment to resolve environmental problems. A variety of regulatory instruments lie in between (Hanke and Boland, 1972).

General Classes of Policy Instrument

1. Rules affecting private contracts and litigation;

2. Regulatory instruments:

    a. Direct subsidies;

    b. Effluent standards, moral suasion, permits (no-cost
        permits a la "1899" model or user charges),
        sanctions, fines, and effluent charges;

    c. Prohibitions;

3. Direct government investment.

2.2.4 Rules Affecting Private Contracts and Litigation

Rules affecting private contracts and litigation can be effectively used to facilitate the internalization of external costs. For example, even when externalities do exist, no misallocation of resources will occur if the parties involved are in a position to negotiate to their mutual advantage with no transactions costs (Coase, 1960). Negotiation will lead to an argument either for the party imposing the damage to compensate the damaged party or for the potential damaged party to pay the potential damager to refrain from the activity which generates the externality. Therefore, policies which reduce transaction costs will facilitate private negotiation and reduce the effects of externalities.

When the use of common property resources is at issue, the costs of negotiation are usually far from zero, and are likely to be prohibitive because of requirements for getting large numbers of people together and for avoiding freeloaders. However, even if private negotiations to reduce externalities fail, there are several common law courses of action available. These include: public and private nuisance, negligence, trespass, strict liability, products liability, and liability for hazardous substances.

Private litigation focuses upon particular disputes between particular parties, not upon broad social policy. This is a shortcoming of using the courts to bring about an environmental change. There are additional difficulties in using the courts, especially in the use of common law actions. These include: the high costs of litigation, locating competent counsel, the persistence required of the plaintiff who may face months or years of lengthy appellate court review, and bringing the law suit in an attentive and able court. But the problems of a common law action, whether it is a nuisance action or any of the others mentioned above, are largely built into the nature of the actions themselves. The larger the number of both plaintiffs and defendants in a single action, the assessment of any one of the plaintiff's share of damages recovered or any one of the defendant's fair contribution to total damages generated, the difficulty of showing that the pollution actually caused the harm shown, the challenge of putting a money value on losses that plaintiffs suffer from pollution, and the weighting of burdens of proof and presumptions in favor of environmental defendants, are all problems which make common law private legal actions difficult to use in improving the environment.

Recommendation

We recommend that efforts be made to lower transaction costs associated with negotiating private contracts and initiating private litigation.

2.2.5 Direct Subsidies

Subsidies usually carry with them the liability of misallocating resources by providing decision makers with artificial signals. The trouble is that subsidies substitute the judgment of outsiders for that of the polluter on what is the cheapest way of reducing emission (Kneese and Bower, 1968). Because it can be granted only for something that obviously appears to be pollution reduction equipment, it biases pollution reduction actions toward those methods that can qualify, to the neglect of other ways of reducing emission that may be much cheaper to society. Thus, pollution reduction procedures that do not qualify for the preferential treatment, such as process change or input change, tend not to be used even though they may be the cheapest methods of reducing pollutant emissions (Plourde, 1972).

If subsidies are granted for the construction of municipal sewage plants, where is the stimulus to manufacturing or other plants to make process changes or changes in composition of inputs that would achieve the same reduction in emission at lower cost? A charge to the private plant for use of municipal sewage processing services would remedy this difficulty if set at the right level. Under the 1972 Amendments to the Federal Water Pollution Control Act, grantees are required to impose charges sufficient to recover the federally financed portion of the extra cost occasioned. It remains to be seen whether such charges will be imposed effectively.

Recommendation

We recommend that the direct and indirect costs of subsidies be carefully studied and explicitly considered when policy changes are considered.

2.2.6 Effluent (Emission) Standards

An effluent standard places a limit of some sort on the effluents
permitted from a source. Such a limit may be expressed in various
ways and, of course, may be quite complex. It might require that emission
not exceed a certain percentage of the offending substance that is
present in plant inputs (e.g., 90 percent elimination), or it might
be expressed as a maximum concentration in stack gas. The standard
may also vary with location and weather conditions in recognition of
the fact that ambient concentrations, which are what the damaged
parties are concerned with, depend not on emissions alone but also
on location and weather conditions.

      Must effluent standards be uniform as among firms and localities?
Clearly these types of uniformity are not a necessary feature of the
effluent standard device, but the tendency toward uniformity in
application is strong, since uniformity has a superficial and politically desirable appearance of fairness about it. Also, it is easier
to administer than differential standards. But such uniformity may
be quite uneconomic. (Johnson, 1967; Porges, 1969; Schultz and
Gutmanis, 1972.) If a number of plants in a vicinity are to reduce
their emissions of, say, particulates to a level that would insure a
certain ambient standard, it would be desirable that the reduction
take place in those plants that find it cheapest to do so. Similarly,
it would be rational that effluent standards in some areas should be
far more stringent than in others. Again, the tendency toward uniformity of an effluent standard works against differentiation based
upon varied costs and benefits.

      The significance of effluent standards can be rather clouded by
provisions requiring that economic and social costs and benefits be
taken into account in determining whether the effluent standards can
be implemented with available technology or other alternative control
strategy. The "Clean Water" legislation of 1972, for example, pays
little attention to the problem of getting pollution reduction at
the least possible cost. Not only does it require effluent standards to
be met in ways of dubious enforceability, but a new over-riding criterion
is introduced, namely, the "best technology." By 1983, both communities
and industries would be required to apply the "best available technology."
The enforcement mechanism is based on a determination of whether or
not the effluent in question is up to the emission standards made
possible by the best available or practicable technology. (See Chapter 3.)

      Are these criteria really definable in practice? They may be in the
field of municipal treatment of domestic wastes, but in most of the

complex industries, such as the petrochemical, metallurgical, or
several of the biological industries--where reduction of pollution is
not confined to waste treatment but also consists of complex processes
within the plant, it may be asking too much of government agencies to
know as much about industrial technology as industry itself knows.
The legislation assumes that EPA can call the shots in this technical
game, but it seems more reasonable to expect that EPA could be outmaneu-
vered by superior industrial knowledge in many cases, with the result
that progress toward higher levels of water quality might be frustrated.

One of the difficulties if the standards are uniform is that they
do not permit some areas to be more polluted than others except insofar as the standard does not act as a constraint simply because emissions
are not sufficient to push against the standard. It would seem rational
that in deciding the kind and quantity of pollutants to be allowed
in an area, account should be taken of the costs of reducing emissions
and also of the damages caused.

Recommendation

We recommend that the drawbacks of effluent standards, which are now
widely used policy measures to deal with environmental disorders, be
carefully studied, and that these drawbacks be explicitly addressed
when policy changes are considered.

2.2.7 Design Standards

Design standards refer to equipment or processes designed to achieve
given effluent standards. The "best practicable or best available
technology" requirement of the Clean Water legislation represents an
equipment design standard, although what is "best" depends on the
objectives one is trying to attain. A design standard could apply
to the more readily locatable large-scale point-sources of emissions,
such as a smelter or to any so-called "non-point" source, such as the
automobile. Thus, it may well be that the preferable method of control
is to regulate the design of original equipment and to require that
its maintenance cost be very low for some specified mileage or period--
as indeed we are attempting to do. A system of design standards must
be developed very carefully, however, in order to avoid the imposition
of unnecessary costs in the attainment of given amounts of reduction
in environmental disruption.

Recommendation

We recommend the careful consideration of design standards, particularly
in those cases in which the measurement of emissions is prohibitively
expensive.

2.2.8 Permits

The fundamental question with the permit device is: what are the

criteria on which a permit is to be issued to a firm? Once a permit
is issued, does it convey the privilege of discharging whatever the
firm wishes or are there limitations? If there are limitations, what
are they and how are they determined? Actually, there are <u>no criteria</u>
under the present procedure to govern the quantity or composition of
effluents by a firm that has secured a permit, as under the Refuse Act
of 1899. Of course, the issuance of permits firm by firm is no bar
to the use of standards that are invariant among firms, but in the
absence of such principles, a firm by firm procedure is a very unsatisfactory policy.

Marketable permits represent another form of instrument that can
be used to control the use of common property resources. Under such
a system, the environmental quality management authority could determine the ambient quality standards and the total allowable effluents
in a particular region, and then auction off permits to discharge
the desired level. The purchasers (dischargers or conservation groups)
could in turn sell or lease their permits. Unfortunately this may only
work in very confined regions (Dales, 1968).

Recommendations

1.   We recommend that the drawbacks of the current permit system
be studied, and that their costs be explicitly considered when policy
changes are considered.

2.   We recommend that experimentation with various types of marketable
permits be started on a regional basis, especially with periodic
re-auctions.

2.2.9 <u>Taxes or Charges on Emissions</u>

In many cases, charges on emissions are the preferred method of securing
a given reduction in pollutant damages (Baumol and Oates, 1971;
Freeman and Havenmen, 1972; Maler, 1972; Upton, 1968). The emissions
charge or tax is specified on each unit of emissions, and is set at
a level that will induce polluters to adjust their discharges so
that ambient quality standards are met. Under such a system, the
dischargers are led to compare the cost of using the environment for
residual discharge (as reflected to them by the emission charges) with
the cost of handling their residual disposal or recycling problems in
some other way. In addition, under the residuals charges strategy
all dischargers face a continuing incentive to economize on the use
of the environment by further reducing discharges, since they are
being required to pay for the right to use common property.

With a system of charges, each firm can make its own decision
on how far it should reduce emissions to escape the tax (charge)
for the use of the environment. With emission standards, outsiders
who cannot know all of the relevant facts, in effect, make these

decisions for the firms. What is more, emissions charges have no
bias toward any particular method of reducing the emission, as is true
of subsidies or with a requirement that certain types of pollution
reduction equipment be installed. The firm is perfectly free to decide
whether it should change its processes, change the nature of its
materials input, initiate treatment at the "end of the pipe," or to
change location, if the charge on effluent varies with location.
In addition, the tax will cause the price of the article to increase
and consumers will then have to decide the extent to which they wish
to reduce consumption of the article in view of its higher price.
In some cases, it may turn out that the easiest form of adjustment
to secure emission reduction is by having the consumer change the
composition of his expenditures. This effect is present, of course,
with the use of standards. That is, if it is very expensive to comply
with the emission standards, the price of the commodity will increase
with a corresponding consumer response. The point is that the desired
reduction--whatever it is--will be secured more economically with,
rather than without, charges.

   Charges on emissions of effluents do _not_ represent a license to
pollute, but have an entirely different effect. Since the polluter
must pay for each unit of pollutant he emits, he finds that he asks
himself how far it will pay to _reduce_ emissions in order to avoid
paying the tax. He will reduce emissions up to the point where the
cost of further reduction is greater than the tax on a unit of
pollutant emitted.

   There are problems associated with using a system of effluent
charges, but it is important to distinguish between problems that are
common to both standards and charges and those that are peculiar to
the use of charges. There are circumstances in which some type of
effluent, or design standard may be superior to imposition of an
effluent charge. Quite often, however, the alleged superiority of a
standard as a form of regulation is based on a fallacious argument.

   For example, it is often said (correctly) that a serious difficulty in designing an effluent tax or system of taxes is that quantitative estimates of the benefits to be had from pollutant reduction
are difficult or impossible to make. If one then concludes that a
standard is superior, he has made a logical error. Presumably the
reason for imposing a standard is because we expect the benefits to
outweigh the costs of meeting the standard. Thus the problem of making
up our minds on the prospective benefits is _not_ avoided by choosing
to use standards as a regulatory device rather than an emission tax.
After all, the same benefits are involved in both cases! However, in
many cases, an effluent charge system would be the least costly way of
obtaining these benefits. In addition, the proceeds from the emissions
tax might be used to acquire the knowledge needed on controls, substitutes, locations, options for technological change and other relevant
facts as well as aid in financing direct government investment in
environmental disruption abatement (see Section 2.2.10). One case in

which a certain type of standard may be superior to a charge is that
in which measurement of effluent would be difficult or expensive, such
as with automobile vehicle emissions. (See the above discussion on
automobiles and design standards in Section 2.2.7.) In the case where
the desired rate of pollutant emission is zero, as when a substance is
so dangerous that the preferred course of action is to avoid its use
entirely, there is nothing superior in the use of a tax set so high
that emissions will be cut to zero. A flat prohibition secures the
same result.

The superiority of the charge system arises when economizing
adjustments are possible in securing a given reduction in emissions.

Recommendation

We recommend that in most cases where the policy goal is to minimize
the cost of pollution abatement plus damages from emissions, an
effluent charge be the preferred policy device, other things being equal.

2.2.10 Direct Government Investment

The last major strategy for dealing with externalities is the
practice of minimizing the externality through direct government
investment. In this case, the government assumes the responsibility
for reducing the impact of side effects. An important concept in this
discussion is that of the returns to scale associated with the technology
required to accomplish the reduction. In the case of increasing
returns to scale (e.g., the advantage of building one large regional
waste treatment plant rather than a number of small, less efficient
ones, the advantage of governmental reclamation of abandoned surface
mined lands, or the advantage of governmental in-stream aeration of
streams where pollutants have used up available oxygen supplies),
an excellent argument for the efficiency of direct government investment
may exist (Davis, 1968). By combining the effects of a number of firms
and reducing them all through a single action, the government may realize
significant cost savings for society as a whole. Should the technology
be characterized by constant or decreasing returns to scale, however,
these savings will not exist, and the action of government in externality reduction enjoys no advantage over the same action taken by individual firms.

Recommendation

We recommend that direct government investment be considered when
economies of scale characterize the technology available for the
amelioration of a large-scale, wide-spread environmental problem.

2.2.11 The Planning Stance

Many environmental problems are so complex that comparatively simple devices, such as emission charges or effluent standards, may not be usable without considerable elaboration. In the case of strip mining, for example, it is difficult to define some of the externalities. There may be several adverse environmental effects that cannot be controlled in any simple way.

Cases of conflicts in the use of resources, particularly where aesthetic effects are involved, are frequently very difficult to handle. Examples include mining vs. recreational use of neighboring lands and locational decisions which transfer the geographic impact of environmental disruption. Conflict between operations involving petroleum-- shipping, unloading, and off-shore drilling--and aesthetic appreciation and recreational use of the shore are becoming more frequent and severe. In many cases involving highways and other aspects of urban and suburban design, the externalities cannot be fully measured or even fully described.

There certainly are no ready solutions for all of these problems, although some are much less intractable than others. Examination of our recent history suggests that single purpose planning is unlikely to be successful, even for the purpose for which it is conceived, unless it is coordinated with planning for other closely related goals and includes planning for the other closely linked subcomponents of the major system. Environmental planning should thus be heavily weighted in all national, regional and urban planning activities for land use, urban and regional development, and other activities identified above-- a principle already recognized in the National Environmental Policy Act, requiring the preparation of environmental impact statements for all major programs and projects undertaken by the federal government.

Recommendation

We recommend that comprehensive, integrated planning involving assessment and public disclosure of anticipated environmental impacts be required for significant activities.

2.3   Evaluation Criteria and Tradeoffs in the Application of Policies
        to Certain Environmental Problems

Numerous evaluation criteria and tradeoffs must be considered by policymakers and managers of economic units in order to resolve or ameliorate critical environmental problems associated with the materials economy. The present section addresses some of the more important factors.

2.3.1   The Location Element, Differential Assimilative Capacity, and
        Land Use Policy

The location element enters environmental problems in many guises. The way it is handled has a profound effect on the production and use of materials.

In some cases, the locations of emitters and receptors is not a matter of much importance, either because damages are not altered by changes in the locational configuration or because geographical considerations so constrain the location of emitters that the only locational adjustment that is possible is on the recipient side. A substance like DDT, whose local concentrations certainly vary with differential rates of application but which also is transported through air, water, and the food chain to widely dispersed places, is a case in point--it probably would not be worthwhile trying to differentiate the charges by location of application. Sulfur dioxide (SO_2) certainly is not in the same category in this respect, but atmospheric sulfur compounds are transported substantial distances. Thus if the number of emitting points becomes large enough, we may not be justified in ignoring the locational aspect completely, but it may be necessary only to consider regional differences without trying to make more precise local distinctions. On the other hand, particulate emissions may affect visibility over very large areas, an effect that may even concern people who spend only a small portion of their time in the affected area, and differentiation of charges (or of standards) by location may not achieve a great deal. Differential tax charges (or standards) would be justified only if it were possible to perceive a differential locational configuration that would represent a genuine improvement at the national level.

However, location _does_ make a difference in many and perhaps most cases of pollution. Until now problems of pollution control seem to have been considered largely on the assumption that the locations of both polluters and damaged entities are fixed. In the short run this is the correct assumption, of course, since it is not possible to change very rapidly the location pattern of industrial plants or even of residential areas. Even in the short run, however, rational control policy would not ignore locational considerations since a given rate of emission may produce widely differing damage rates depending on the relative locations of potential recipients of the pollutant and the transport mechanism by which the pollutant is removed. In spite of the clear importance of the location element even in the short run, our present regulation systems by and large do not pay much attention to the location factor.

Even less attention is paid to the longer run problem. It would, for example, be desirable that charges on pollutant emissions or emission standards vary by location so as to encourage a shift in the sites of pollutant sources to those places that are most advantageous for minimizing the combined costs of pollution control and damages from pollutants. Merely to pose the problem in this way should serve to remind us that there is at present a notable lack of coordination between efforts to control pollution and the machinery by which land use is determined and regulated. There surely is an opportunity here for the development at all levels of government of a much better integrated structure of policies for environmental quality, materials use, and land use than we have at present.

Recommendation

We recommend that locational elements which affect environmental quality be given special consideration in materials policy so that selectivity can be applied nationally in the use of lands where materials are obtained, processed, used, and disposed. Selectivity is essential so that the combined costs of pollution control and environmental disruption are minimized.

2.3.2   Specific Locational Environmental Problems: A Special Case

A few locational problems are worthy of specific mention because they have some important features in common. The past several years have seen many controversies involving the question of mining activities in wilderness areas or areas that may at some time be included in the wilderness area system. The Wilderness Act of 1964 sets up a group of procedures which presumably ought to resolve these conflicts, but it is only too clear that they are insufficient to do so.

In some parts of the country, a great debate is going on about the extent to which strip mining should be allowed, one position being that it should be prohibited completely. In the case of Alaska, the lines between development and preservation of assets as they are is presented in a more generalized form. On the one hand, there are gigantic, comparatively untouched natural features-- such as the Brooks Range--which could be preserved without development. On the other hand, we have a state government, with the property interests behind it that are behind all state governments, intent on investment which will have the effect of increasing the market value of their properties.

How are these conflicts to be resolved? The final answer to this question cannot be given here, but a few features of the problem that seem not to have received sufficiently wide recognition may be discussed.

One feature which seems to characterize these problems is a greater polarization of demands than is true of the "ordinary" pollution problems. Perhaps what is involved is the feeling that if the few remaining comparatively untouched natural areas vanish, there really is nothing to replace them, whereas pollution in urban areas is something that one can avoid if one chooses. Whatever the proper explanation may be, greater polarization of demands does seem to be a fact. It may well be that the procedures we have adopted to resolve these conflicts contribute to the polarization of demands.

A possible hypothesis to account for this is as follows: suppose opinions in the country are divided in such a way that if we were to take a vote on each of fifty forthcoming cases of conflict between preservation and mining activity, that in each case the vote would be

52 percent in favor of mining activity and 48 percent in favor of
wilderness preservation. Now if the procedure for deciding these
conflicts did in fact involve a case by case decision on the basis of
a vote, every case would be decided in favor of mining activity and
those on the losing side would complain, with considerable justice,
that they were suffering the tyranny of the majority.

But there may be a better way. Suppose we decided we would like
to simulate what happens in the competitive market. What this would
mean in the case of the wilderness areas in conflict would be to try
to divide them up so as to reflect at least roughly the intensity of
the demands for mineral products. That is to say, if mineral products
can be had from other locations, mining companies would find it
difficult to outbid the wilderness lovers if a bidding process actually
were in effect. (See section 2.2.8 on marketable permits.)

If our decision procedures had this type of allocation as the
objective, with an attempt to take into account future as well as
current demands, perhaps each side would feel more comfortable.
That is, instead of a series of independent determinations, we
need to make these determinations against a more general picture of
how the two competing demands are going to develop through time.

Recommendations

1.   We recommend that experimentation begin on bidding systems that
would allow various interest groups to reflect the intensities of
their demands for various areas.

2.   We recommend that systems of conditional leasing be developed
which would protect environmental values by requiring bonding and
acceptable extraction and reclamation practices.

2.3.3 Intangible Benefits

What should be done with intangible benefits of pollution control which
may be hard to measure? Should they be included in benefit-cost
calculations or not? Is the destruction of a scenic view a real
cost or not? Clearly it is. If a tangible activity constitutes a
legitimate demand on resources, so do intangible benefits.

Many intangible benefits are directly connected with goods that
are bought and sold. These create no problem for benefit-cost
analysis. Even in those cases where this direct connection does not
now exist, it may be possible to make some indirect inferences about
the money values that people are attaching to them by inferences from
observable aspects of their behavior. For example, there are no
market price data to reveal directly the demand for certain types
of outdoor recreation. Since the persons using a given site come from
widely different distances, however, it has proven possible to infer
how much they would be willing to pay for this recreation by careful

analysis of the economic factors, such as travel cost and time, involved in the decisions to go to the site. Another method that has been used to derive information on demand is the questionnaire (Knetsch and Davis, 1966). If carefully constructed and analyzed, very useful information on the value of intangible benefits to the persons experiencing them can be assembled.

Unfortunately these methods cannot take care of all problems associated with intangible benefits. The most intractable problems of all involve cases where the intangible benefit represents a very large part of a person's psychic income. It is not hard to imagine cases involving sums of money that are very large relative to the outlays that would be required to avoid discharging particulates and other pollutants into the air. Reasonable though that possibility may seem, no one has yet succeeded in systematically estimating this value. Thus, because of measurement, aggregation, and non-comparability problems, there is no escape from discussion and use of the political process to try to reach some consensus on the size of the group adversely affected and on the strength of their feelings.

In cases like this, the net sum of benefits and costs will have to be calculated without including the hard-to-value intangible items in the total. A negative sum in this case would not necessarily indicate that the project is undesirable, but rather that the hard-to-measure intangible benefits, if they are to tip the balance in favor of the project, must be worth at least this amount to somebody. The only way to decide this is the political process.

It should be noted that some of the goods and services involved in benefit-cost calculations have the property that more than one consumer may "consume" them simultaneously without any diminution in the quantity of the service available for consumption. Scenic views are of this nature, as are also radio and television signals, and also many pollutants. With respect to the last, if one million people in an area are "consuming" the SO_2 in the area, this hardly diminishes the quantity that can be "enjoyed" by another person who comes into the area. The practical significance of this property, which is not possessed by ordinary goods like bread which can be consumed by only one person, is that in totaling the benefits (or costs) we must add up all the benefits enjoyed by all the persons involved. If SO_2 in the atmosphere is reduced, ideally we should add up the benefits accruing to all the persons involved. Thus a small benefit enjoyed by each of several million people in a region may add up to a very large total and justify a sizable cost to attain it.

Recommendations

1. We contend that intangible benefits, even though difficult to measure, are real effects, and recommend that they should be part of an evaluation procedure which includes benefit-cost analysis of various policy alternatives.

2.   We recommend that in those cases in which intangible benefits
cannot be measured, the measurable benefits and costs of various
alternatives should be calculated, facilitating the political process
that must ultimately resolve the valuation problems associated with
intangibles.

2.3.4 Primary Production and Recycling

Just why recycling has such widespread appeal as an apparent remedy
to many environmental problems is not completely clear. There are
many possible connections. If you re-use a pound of copper, environmental
costs associated with the site of primary production of _copper_ will
have been avoided to that extent. Unfortunately, there may be other
environmental costs associated with secondary production that are just
as important. If a container is re-used, it cannot be deposited
into the environment as litter. Many seem to believe, however, that
recycling usually results in a net savings of inputs, probably reason-
ing from household or personal experience with "using things over again."

However, there can be no general presumption that recycling is
superior to primary production. The person who entertains such a view
has the unenviable task of explaining why only limited quantities
of many materials are recycled. The economizing problem in every
case is one of attaining that degree of recycling which pays off
for society.

If more inputs are used up to assemble articles containing
copper, to transport them, to take them apart, and to melt them down
than are required to produce the same quantity of primary copper,
why should this particular act of recycling be undertaken? On the
other hand, much recycling goes on because it pays firms to recycle.
With each material, however, there are many discarded articles whose
collection and processing is simply too expensive, again in terms of
the real resources required to do so.

The real question for policy is whether there are aspects of
existing policy that distort the balance between primary and secondary
materials, and the chances are that this is in fact the case--with the
balance tipped in favor of primary materials.

One source of bias in policy is quite clear, namely, primary
production receives favored tax treatment in the form of the complex
package involving percentage depletion, expensing privileges, and
opportunities to convert what otherwise would be ordinary income into
capital gains. Measured at the level of the material itself, the privilege is substantial in relation to price, as is more or less obvious
from the fact that percentage depletion is applied to the first saleable product from the extraction activity. This would be at the concentrate
level for many minerals. In addition, the primary industry receives
something of a subsidy from the provision of certain geological and
other mapping services by governmental agencies. Tax provisions
in some states no doubt serve to tip the balance further in the same
direction.

Certain transportation rates tend to favor primary production as against secondary (e.g., rates for concentrates of ore vs. rates for scrap). No easy generalization is possible here, for the situation varies from case to case. In addition, whether there is economically significant discrimination is not a simple question.

The erasing of the differential tax treatment might take either of two routes: granting additional privileges to secondary recovery activities; or eliminating the favorable position granted to primary production. It is fair to say that most economists who have studied these problems, including those who are especially expert in taxation, take the position that the special tax treatment accorded primary production as against other business activity ought to be eliminated. Tax privileges encourage more investment in the favored industries and a (subsidized) lower price than otherwise would be the case. Primary materials tend to be cheaper than they should be in relation to other goods and services. Elimination of the privilege would automatically secure equality of treatment as between primary and secondary production. To go the other route--that is, to extend the special tax privileges to secondary production--would be to make an already complex tax system even more complex.

The container problem and the recycling issue in general involve, in some cases, the additional element of external costs associated with disposal of the original container, an aesthetic cost in this case. It is worth something to reduce these costs and this ought to be an element in any decision on schemes designed to encourage or compel the re-use of containers. Once again, careful cost calculations are essential. The result is not necessarily the same for different types of containers nor for different localities.

A tax-rebate device could be applied even to this case. Actually, a number of variants have been proposed, the common ingredient being a charge at the time of purchase that is refunded on return of the containers. Some consumers could not be bothered, of course, but the opportunity for gain might induce others to pick up their discards.

Recommendation

We recommend that the biases in current policies that favor primary over secondary materials be eliminated. These include such things as differential transportation charges and tax provisions (percentage depletion allowances, expensing privileges and some capital gains provisions).

2.3.5 Technological Change and Large System Effects

The relations between technological change, environmental problems and materials production and use are very complex and cannot be described

in any simple way. (Ayres and Gutmanis, 1972; Kneese and Bower, 1968).
Among the effects of technological change are the following:

1. Increase of market-connected output per unit of input
    (as measured by market prices);

2. Materials consumption--effect may be in either direction;
    it may be positive for some, negative for others;

3. External costs--no generalization possible. Some may be
    increased, some decreased.

What has been responsible for the increased severity of
environmental problems--increase of population, increase of output
per capita, or technological progress? The answer to this question
is not in yet, but whatever it turns out to be, it no doubt will
vary from problem to problem or, viewed from the other angle, from
one technological development to another. Interaction among the
three variables may be required for solution of most environmental
problems. A freeze on any of the variables is not enough. Neither
can technological fixes provide all the answers.

Whatever the net effect of technological progress has been on
our environmental difficulties, certainly the gross effects have
been of great importance, easing our problems in some cases and exacerbating them in others.

It is important to understand how ill effects may arise from
technological progress. The key to understanding is the idea that
external effects have not been adequately taken into account in
making decisions about production and consumption. Suppose that when
a new process is introduced all of the costs associated with it
are reflected in the costs of the firms using it; for example, health
effects arising from stack discharges or the additional contribution
to particulate air pollution. In such a well-ordered world, we could
be sure that the net effects of the introduction of new processes and
products would be positive. There might be adverse effects, of course,
but they would be more than balanced by desirable effects.

We are not ever likely to have this well-ordered system, but it
should be noted that as machinery and procedures are developed that
force decision makers to take into account a larger range of effects,
we shall probably move closer to that position. The chances should be
increased that technological changes actually put into operation will
have beneficial effects. One important example to note here is that
charges for emission of pollutants continue to provide a reason for
the emitter to find a way to cut his emission even more in order to avoid
the charge. Contrast a once and for all permit--once you have secured
the permit, why exert yourself to cut down emission of the pollutant?
Or if there is an emission standard, there is no reward to be had
from reducing the emission below the standard.

The basic uncertainty in the question is whether important effects flowing from change are not recognized or captured in our regulatory systems or, if recognized, are not dealt with effectively. It is quite conceivable that what we might call large system effects such as these could cause us much difficulty in the future even though we are quite successful in learning how to cope with apparently simpler and more straightforward pollution problems.

Although large system effects may be a source of much difficulty for us, it is also quite possible that there are large system re-organizations that could improve our position markedly with respect to environmental problems but which have no chance of adoption because our institutions are not geared to making or permitting large discrete changes.

One possibility, on which considerable work has already been done, is to use smaller but more comprehensive energy complexes such that the heat from electric power production that is now wasted could be used for space heating and perhaps some processes. The elements involved in the system evaluation are not obscure. They involve economies of scale for different types of generating equipment, different fuel and fuel transport costs, differential transmission costs, and, of course, the problem of distributing electrical energy to those consuming points that are not dense enough to warrant the same kind of "efficient" energy complex but which would have to be located either with reference to a conventional distribution system or would have to suffer the cost of even smaller scale production of electrical power. The result of a thorough study of the possibilities might well be that the energy complex idea is advantageous for densely populated or industrialized areas, but that conventional production and distribution is better for other areas.

Suppose these indeed turned out to be the results. It is easy to see that so fundamental an innovation in the organization and supply of electric power would have a number of very difficult hurdles to cover, most of them stemming from the fact that the production and distribution of electric power is highly regulated and that there already is a system in operation whose phasing out would have to be carefully integrated with the introduction of the new system.

Another potential large system change that might be advantageous in many ways and reduce a number of environmental impacts is compre-hensive change in the configuration of urban activities and transport. At least in part, it is clear that we have the commuting and rush hour traffic problem because people do not have available the opportunity of purchasing an arrangement without it that does not at the same time involve costs too great to be acceptable.

It is apparent that large system modifications are a very difficult

thing to handle. On the one hand, it _is_ difficult to be sure that
the large-scale modification has been properly costed out, to be
sure that all of the effects have indeed been foreseen. The hesitation
in supporting proposals for large-scale modifications is probably
very well justified. We have only to think of the many proposals
that have been made which have been shown in a very convincing way
to have serious defects--fortunately before they were instituted.
In other cases, the plan has indeed been sold and bitter experience
has clearly indicated costly effects that the planner was not aware of.

On the other hand, the rewards to be had from designing a good
large-scale modification and figuring out how to put it into effect
may be very great. One possible way out of this dilemma is to experi-
ment on a scale large enough to develop and bring into the open the
effects of the proposal not anticipated by its designers, but still
on a scale much smaller than wholesale adoption of the proposal would
involve. It would hardly be necessary to commit ourselves to modification of the electric power production and distribution system of
the whole country in order to get a good idea of whether the self-
contained complex idea is viable or not.

Recommendation

We recommend that national policy should promote the estimation
of large system effects through experiments on a scale large enough to
develop and reveal the effects of a proposal not anticipated by its
designers but on a scale much smaller than general adoption.

2.4   Should the Overall Rate of Economic Growth Be Reduced?

A proposal to cut economic growth in general could be justified only
on the ground that failure to do so would involve us in penalties so
great as to be recognized as too great by almost everybody. To argue
otherwise leads to another problem--that of arbitrating among different
tastes. It may be that we are in fact at the point of suffering
substantial penalties, but it is only fair to say that no one has yet
demonstrated this in any convincing way that goes beyond the expression
of his own tastes for a less frenetic, a less materialistic society.

That would, indeed, be a simple resolution of the problem.
Actually, it seems at this point that our problems are more subtle
and probably would continue to plague us even if we were to restrict
the overall rate of growth. Damage functions, that is, the damages
to be expected from various levels of emissions of pollutants,
are in fact highly variable, so that it could easily be that a general
restriction in growth would still leave us consuming too much of some
substances. Our society is not likely to accept or accede to a general
limitation on growth whereas it may approve limitations on the emissions

of particular substances where there has been some showing of subtle damage that cannot be perceived by direct observation or where the offense is directly perceptible, as it certainly is with visible air pollution, for example.

It is important to realize that there are various routes to a general reduction in the rate of economic growth, say to zero. One possible route is to decide, suddenly, that we should go to a rate of zero growth. But suppose we follow the route of restricting the use of substances that exhibit damaging effects in one way or another. What will be the effect on economic growth as a whole? There are various possibilities. It may be that some important substance comes to exhibit adverse effects so serious that its use must be curtailed and that its curtailment will bring the whole process of expansion pretty much to a halt. What could this substance be? We might take a look at the energy commodities. There are several possibilities here, including upset of the heat balance, adverse effects of oil spills on ocean and stream life, inability to control radioactive substances emitted in the course of producing electric power from atomic fission, etc. Or we might conceive of a cumulation of adverse effects flowing from increased consumption (and disposition) of substances already in use plus the burden of new substances increasing to the point where growth surely is slowed down and perhaps stopped altogether.

We should be alert to the possibility that the brake on growth may come, not from the adverse effects flowing from the consumption of certain materials, but from the adverse effects of congestion. These conceivably could lead to very strong pressures to contain economic growth.

2.4.1 Uncertainty

The question of whether the overall rate of growth should be reduced is closely tied up with the question of uncertainty. One of the most powerful arguments in favor of a general restriction is that our inability to perceive ill effects far enough in advance or clearly enough exposes us to the danger of crossing one or even more points of no return and waking up to find ourselves in a very undesirable position without being able to do anything about it.

That these dangers are serious is conceded--or insisted on-- by many people. Unfortunately, economics has no surefire prescription for such situations. Its advice tends to be rather obvious and common-sense.

Several courses of action ought to be followed simultaneously. First, we need to extend our understanding of ill effects associated with the use and disposition of the thousands of different compounds to which we are exposed through a multitude of avenues. This is a difficult thing to do since some of the effects may be years in the making, as with certain carcinogens, and others may be the result of the

combined effects of exposures to many different substances. These
remarks apply to effects on the human body and psyche, and also
to effects on the many different species found in the various natural
environments.

Second, we need to extend and increase the amount of systematic
observation of natural systems so that we may get as early warning
as possible that something is changing even though we may not at
the moment understand what is causing the change or how important
it is. Recognition of the fact of change may suggest ways to investi-
gate and search for the possible causes. Recall that rather casual
observation of changes in the population of robins in various places
a few years ago played an important role in reaching a consensus on
the place that one pesticide should have in our society.

Finally, if we are running the risk of going beyond points of
no return, common sense certainly urges that it is better to err on
the side of caution. It would not seem to be wise to place too much
of a burden of proof on those who contend that use of a certain
substance has harmful effects. The object of the game is <u>not</u> to
completely avoid needless restrictions in use or consumption but
to avoid those failures to restrict that will involve us in major or
even catastrophic penalties.

Recommendations

1. We conclude that the type of growth (the mix of final goods and
services) is the important issue that must be faced, not the "no
growth" option.

2. We recommend that attention be given to the following considerations, particularly in cases where uncertainty prevails:
the systematic observation of complex natural systems so that we may
obtain as early a warning as possible that something is changing even
though we may not understand what is causing the change or how important it is.

References

Air Pollution Control Administration (APCA), 1969. _Air Quality Criteria for Particulate Matter_. (Durham, Research Triangle Park: Publication AP-49, APCA), 211 pp.

Air Pollution Control Administration (APCA), 1970. _Air Quality Criteria for Carbon Monoxide_. (Durham, Research Triangle Park: Publication AP-62, APCA)

Air Pollution Control Administration (APCA), 1970. _Air Quality Criteria for Hydro-Carbons_. (Durham, Research Triangle Park: Publication AP-64, APCA)

Air Pollution Control Administration (APCA), 1970. _Air Quality Criteria for Photochemical Oxidants_. (Durham, Research Triangle Park: Publication AP-63, APCA)

Air Pollution Control Administration (APCA), 1970. _Air Quality Criteria for Sulfur Oxides_. (Durham, Research Triangle Park: Publication AP-50, APCA), 178 pp.

Air Pollution Control Administration (APCA), 1971. _Air Quality Criteria for Nitrogen Oxides_. (Durham, Research Triangle Park: Publication AP-84, APCA), 175 pp.

Ayres, R., and I. Gutmanis, 1972. "Technological Change, Pollution and Treatment Cost Coefficients in Input-Output Analysis," _Population, Resources, and the Environment_, R. G. Ridker, ed. (Washington, D.C.: U.S. Government Printing Office), pp. 313-337

Bator, F., 1958. "Anatomy of Market Failure," _Quarterly Journal of Economics_, vol. 72, n. 3, pp. 351-379

Baumol, W., and W. Oates, 1971. "The Use of Standards and Prices for Protection of the Environment," _The Swedish Journal of Economics_, vol. 73, n. 1, pp. 42-54

Caldwell, L. K., 1963. "Environment: A New Focus for Public Policy?" _Public Administration Review_, vol. 23, n. 3, pp. 132-139

Coase, R., 1960. "The Problem of Social Cost," _Journal of Law and Economics_, vol. 3, pp. 1-44

Dales, J. H., 1968. _Pollution, Property and Prices_. (Toronto: University of Toronto Press), 111 pp.

Davis, R., 1968. _The Range of Choice in Water Management: A Study of Dissolved Oxygen in the Potomac Estuary_. (Baltimore: Johns Hopkins Press), 196 pp.

Demsetz, H., 1964. "The Exchange and Enforcement of Property Rights," Journal of Law and Economics, vol. 7, pp. 11-26

Federal Water Pollution Control Administration (FWPCA), 1972. Federal Water Pollution Control Act Amendments of 1972. (Washington, D.C.: U.S. Government Printing Office), 1 p.

Freeman, A. M., and R. Haveman, 1972. "Residual Charges for Pollution Control: A Policy Evaluation," Science, vol. 177, n. 4046, pp. 322-9

Hanke, S. and J. Boland, 1971. "Water Requirements or Water Demands?" Journal of the American Water Works Association, vol. 63, n. 11, pp. 677-681

Hanke, S. and J. Boland, 1972. "Thermal Discharges and Public Policy Alternatives," Water Resources Bulletin, vol. 8, n. 3, pp. 446-458

Hufschmidt, M. M., 1971. "Environmental Quality as a Policy and Planning Objective," Journal of the American Institute of Planners, vol. 37,   4, pp. 231-242

Johnson, E. 1967. "A Study in the Economics of Water Quality Management," Water Resources Research, vol. 3, n. 2, pp. 291-305

Kneese, A. V. and O. Herfindahl, 1965. Quality of the Environment. (Washington, D.C.: Resources for the Future, Inc.), 96 pp.

Kneese, A. V., R. U. Ayres, and R. C. d'Arge, 1960. Economics and the Environment: A Materials Balance Approach. (Baltimore: Johns Hopkins Press), 120 pp.

Kneese, A. V., and B. T. Bower, **1968. Managing Water Quality: Economics, Technology, Institutions. (Baltimore: Johns Hopkins Press),** 328 pp.

Knetsch, J. and R. Davis, 1966. "Comparisons of Methods for Recreation Evaluation" in Water Research, A. Kneese and S. Smith, eds. (Baltimore: Johns Hopkins Press), pp. 125-142

Krutilla, J., 1967. "Conservation Reconsidered," American Economic Review, vol. 57, n. 3, pp. 777-86

Maass, A., 1966. "Cost-Benefit Analysis, its Relevance to Public Investment Decisions," Quarterly Journal of Economics, vol. 80, n. 2, pp. 208-26

McKean, R. N., 1958. Efficiency in Government Through Systems Analysis. (New York: Wiley), 336 pp.

Maler, J. C., 1972. "Effluent Charges versus Effluent Standards" (working paper prepared for Resources for the Future)

Margolis, J., 1957. "Secondary Benefits, External Economies and the Justification of Public Investment," Review of Economics and Statistics, vol. 39, n. 3, pp. 284-91

National Academy of Sciences/National Academy of Engineering, 1972. Water Quality Criteria, 1972. (Washington, D. C.: U. S. Government Printing Office), in press

Plourde, C., 1972. "A Model of Waste Accumulation and Disposal," Canadian Journal of Economics, vol. 5, n. 1, pp. 119-25

Porges, R., 1969. "Regional Water Quality Standards," Journal of the Sanitary Engineering Division, Proceedings of the American Society of Civil Engineers, vol. 95, n. SA3, pp. 423-37

President's Science Advisory Committee, 1965. "Restoring the Quality of our Environment" in Report of the Environmental Pollution Panel, (working paper). (The White House)

Public Law 89-234, 1965. Amendments to Federal Water Pollution control Act

Public Law 91-604, 1970. Clean Air Amendments of 1970

Schultze, C. L. and I. Gutmanis, 1972. "The Environment" in Setting National Priorities, the 1973 Budget. (Washington, D. C.: Brookings Institution)

Scientific Committee on Problems of the Environment (SCOPE), 1972. Global Environmental Monitoring, a report submitted to the United Nations Conference on the Human Environment, Stockholm, 1972. (London: SCOPE), 67 pp.

Upton, C., 1968. "Optimal Taxing of Water Pollution," Water Resources Research, vol. 4, n. 4, pp. 865-75

U. S. Department of the Interior, Federal Water Pollution Control Administration, 1968. Water Quality Criteria: Report of the National Technical Advisory Committee to the Secretary of the Interior. (Washington, D. C.: U. S. Government Printing Office), 234 pp.

CHAPTER 3

STUDY TEAM

ON

ENVIRONMENTAL PROBLEMS ASSOCIATED WITH METALLIC AND NONMETALLIC MINERAL RESOURCES

ARNOLD J. SILVERMAN, Department of Geology, University of Montana, Missoula, Chairman

DAVID B. BROOKS, Mineral Resources Branch, Canadian Department of Energy, Mines, and Resources, Ottawa

BERTRAM W. CARNOW, Occupational and Environmental Medicine, University of Illinois Lincoln School of Public Health, Chicago

NEVIN K. HIESTER, Materials Laboratory, Stanford Research Institute, Menlo Park, California

GLENN L. PAULSON, Natural Resources Defense Council, New York, New York

WILLIAM C. PETERS, Department of Mining and Geological Engineering, University of Arizona, Tucson

G. RAYMOND SMITHSON, Jr., Environmental Systems and Processes Section, Battelle Laboratories, Columbus, Ohio

GENEVIEVE ATWOOD, Staff Officer

3.1 Assumptions, General Conclusions, and Priorities

3.1.1 Assumptions

The study team on metallic and nonmetallic mineral resources developed the following brief assumptions to frame its conclusions and recommendations.

Until 2000 A. D. United States population and per capita consumption of material resources will continue to grow, though not necessarily exponentially, despite changing life styles, environmental consciousness of people, and selective, short-term mineral shortages.

The environmental impacts of mineral resource extraction, processing and use can change the total ecological balance of the earth's surface and near surface. In some cases, mineral development will greatly affect the sociological and economic structures of society, but these are not the main considerations of this chapter.

Because of the resulting environmental effects, the study team charge includes the assessment of the impact of the consumption rate on the demand for minerals, but does not focus heavily on the environmental effects of the final consumption of material goods.

The study team did not consider any limited, particular forecast period. However, in many cases it found distinctions between policies appropriate to short- and long-term considerations. For example, differences between proposals for existing mineral resource facilities and proposals for future plants imply such a distinction. Recommendations for system changes such as those possibly required to cope with world-wide health effects, energy policies, and land use planning imply still longer time frameworks.

We focused on problems resulting from normal operating practices and ignored environmental problems stemming from major accidents or natural disasters (such as earthquakes and floods).

3.1.2 General Conclusions

In its considerations, the study team found that it returned to many of the same conclusions pertaining to all or most stages of mineral production.

1. The goal of a national materials policy is to enhance the quality of life by providing for the material needs of people while protecting environmental quality. There is no fundamental requirement to supply a given quantity of any mineral resource; rather, there is a demand for services that such materials can provide, and this demand is a function of economic, institutional as well as technological variables.

2. From the point of view of supplying mineral resources, no
    theoretical _technologic_ limits exist. The worldwide
    availability of most mineral resources is very large re-
    lative to foreseeable consumption rates. For those materials
    that are in short supply, alternate technologies or substitute
    materials will probably emerge under the stimulus of price
    changes. The quantities implied by the word "availability"
    are not just those included in reserve calculations, but
    the much larger quantity in "resources." The environ-
    mental consequences, however, of supplying huge quantities
    of materials to an increasing population are frightening
    and must be regarded as overriding in developing future
    materials policy.

3. The environmental prerequisites for sustained, high quality
    ecosystems remain partially unknown. We do know that to
    maintain and enhance environmental quality in national life,
    perhaps dramatic systems management changes must be developed
    and implemented in waste handling, public health and welfare
    considerations, and business and governmental decision
    making as they impinge upon environmental resources. In
    order to just "maintain" present environmental quality, we
    must expand environmental management (see 5. below), because
    maintenance at status quo will result in a net environmental
    degradation as population and per capita consumption increase.

4. If population and per capita consumption continue to grow,
    even with reduced emissions from mineral resource recovery
    and improved siting practices, environmental problems may
    grow faster than technology can solve them. Therefore, we
    must regard limitations of total demand as one of the more
    critical problems we face as a nation.

5. Much of current attention focuses on pollution, narrowly
    defined, rather than environmental management. With regard
    to the latter, the study team concluded that land use
    planning and management and technologic assessment of the
    materials utilization cycle probably will prove the most
    useful tools for achieving an enhanced environment.

6. Environmental standards will continue to differ considerably
    among and within nations. The desire for a high quality
    environment may be partly sensitive to income levels;
    however, the amount of materials resource development will
    also be sensitive to a variety of ecological factors.
    Therefore, a spectrum of management techniques and environ-
    mental standards will result.

7. The growing affluence and productivity of the individual,
    hence one's higher social value in narrow economic terms,
    increases the benefits derived in higher health levels.

The value of remaining open space, wilderness areas, and unusual landscapes will also increase as demand for recreation goes up and the supply of land space decreases.

8. The United States needs a system of technology assessment to predict or simulate the ecological and social effects of the introduction of new processes and materials. Sometimes this assessment can be accomplished through theoretical evaluation. When this proves inadequate, prototype systems, including all elements of the full system, should be built and operated experimentally on a pilot scale before permitting full-scale operation. Such a step would be appropriate for ocean mining or extensive solution mining of low grade metal deposits.

9. The United States needs a national land use planning act that sets the minimum standards for resource allocation as it effects the environment. Ample opportunity for upgrading the standards should be available to the states.

10. Many of the recommendations stemming from our review imply higher capital costs and/or larger scale operations. Either will tend to promote larger firms or regional cooperation among firms. Therefore, we wish to emphasize the potential anti-competitive effects of some aspects of environmental management.

11. In general, those fields which receive the bulk of research and development funds proceed with technological advancement faster than others. If money and organizational effort are put into research and development for solving environmental problems, the current predictions of large engineering and capital costs to meet high environmental standards may be exaggerated. This does not deny the impact of diminishing returns; it will always be more expensive to recover each additional unit of pollution, but the cost of recovering equivalent levels of pollution will go down.

12. We agree with the President's Council on Environmental Qaulity, and others, that well-planned implementation of pollution control measures will not endanger the viability of either the minerals industry, or of any specific sector within it (CEQ, 1972).

3.1.3 Priorities

The study team considered two broad environmental impacts of mineral resource use that would affect national materials policy. First, we briefly considered the possibility of significant environmental change on a global scale. Certain trends have developed and require watchful concern. These include the increasing acidification of rainfall and total atmospheric sulfation, particularly in the northern hemisphere; an increased amount of heavy metal cycling with food chain involvement; and increasing erosion and sedimentation rates resulting from man's land use activity.

Secondly, we considered the overall level of environmental wellbeing of a population or nation. The following list of environmental priorities are not irrelevant to industrially underdeveloped areas, but we cannot consider them to have highest priority until a population rises above the subsistence level of life (UNCHE, 1971). The following are the environmental priorities considered most important:

1. Human health (both acute and chronic illness) and safety.

2. Environmental and economic health as indicated by other biological and social-psychological parameters.

3. Land use planning concerns in the production and use of mineral resources.

4. Conservation of material and environmental resources through technological assessment and prudent use of substitute and recyclable materials.

3.2 Health Effects

3.2.1 Host-Environment Dynamics and Adaptive Limits

One needs an understanding of some basic concepts regarding environmental impact on human health when considering materials cost, changes in production, need for new technology and tradeoffs or substitutions. Humans as biological organisms with varying degrees of resistance and adaptive capacity continuously struggle with an essentially hostile environment. Anything lowering the resistance of man or increasing the hostility of the environment, decreases his ability to adapt. A number of studies have shown that children and marginal people with poor adaptive ability because of serious lung disease, heart disease, asthma, or other serious chronic diseases, have increased respiratory illness and cardiac and respiratory death rates parallel to the rise in pollution levels (APCA, 1969; CHESS, 1973). Thus, at every level of pollution and not at some defined threshold, depending on the adaptive reserve of individuals, someone becomes sick and someone's life is shortened.

Multiple research efforts considering threshold for a substance frequently have arrived at different results. These discrepancies

undoubtedly relate to variation in the populations under study and
the conditions of life at the time of study. Epidemiological and other
studies strongly suggest that with regard to radioactive substances
and other carcinogens, there is no safe level of exposure or it
may be indeterminably low. Failure to recognize this has resulted
in setting inadequate standards in the workplace and possibly in the
community. Recognition that for some people there is no threshold
will permit defining quantitatively high risk populations at each
level of pollution and facilitate the setting of standards which
protect the maximum number of people.

3.2.2 The Concept of Total Body Burden and Disease

The development and setting of standards in and out of the workplace
has generally been based on considerations of health effects of individual substances from a single source and, as a result, failed
to consider the relation between total body burden and disease.
In considering host-environment dynamics, the total body burden
is a critical factor. Three major areas are involved and all of them
must be considered in defining acceptable levels of pollutants.

First, although standards for a single pollutant may be set for
a single source adequate to prevent most toxic effects, accumulation of
this pollutant from multiple sources may increase the body burden sufficiently to produce negative health effects. For instance, lead enters
the body through ingestion and inhalation, utilizing many transporting
agents including water, food, paint, street dust, volatilized airborne
lead from automobile gasoline, and industrial sources.

Second, in assessing total body burden, the sum total effect of
a number of different toxic substances acting on a single organ
should be considered. For example, in a nation where 60,000,000 people
smoke, we must consider the additive or synergistic effects of this
irritant plus industrial SO_2, and concurrent exposure to NO_2, oxidant,
hydrocarbons, and secondary pollutants in and out of the plant, all
of which target on the lung. Alcohol, solvents, and lead are also
multiple substances that may act synergistically and these target
on the liver. We must recognize the possibility that toxic substances
may enter into chemical reactions that increase the activity of
pollutants. For example, SO_2 oxidation by photochemistry must be
compared to SO_2 plus particulate in the presence of high humidity,
with manganese acting as a catalyst to form sulfuric acid. The latter
changes the response time and intensity of the target organ reaction.

Third, we must consider the impact of other environmental stressors
on the total body burden in addition to those imposed by industrial
and community toxins. Studies suggest that such stressors as low
socioeconomic status with its attendant inadequate medical care,
overcrowding, poor housing, poor heating, and poor nutrition, enhance

the effect of toxic substances by reducing host resistance (Lepper, et al., 1969; Winkelstein, et al., 1967). Workers exposed to the most hazardous substances frequently exist in the lowest socioeconomic group in the industrial process and live closest to the plant. Close scrutiny of this problem should be made. It is as important to lower in- and out-of plant pollutants in the immediate surrounding community, as in the community at large.

3.2.3 The Setting and Achievement of Health Standards

The establishment of health standards in many industrial processes utilizing mineral resources has proved inadequate. There still exists a high rate of lung cancer, mesothelioma, and asbestosis in asbestos workers. Coal miners suffer from pneumoconiosis and associated pulmonary diseases. Lung cancer in coke oven workers has increased noticeably. Part of the difficulty stems from three basic errors in setting standards.

Failure to consider individual adaptive limits results in the use of a survival population to set a "safe level" of exposure for a 40-hour work week. Longitudinal studies strongly suggest that workers in plants using toxic materials, or with high energy demands, are a "survival" population. Those sensitive to toxic materials either die, become disabled and are forced out of the industry, or leave the industry because they find the work unpleasant. "Safe" standards, then, rest upon illness in the most insensitive or strongest population; hence the term "residual" or "survival" population.

2.   Failure to consider total body burden has resulted in standards predicated on a 40-hour work week of exposure to a toxic material. The TLV (threshold limiting value) suggests a concentration and time period over which a worker can be exposed through his working lifetime without suffering ill effects. The TLV does not consider that many workers live close to the plant, and are exposed, albeit at lower levels, to the same toxic materials present in the plant. Examples included SO_2 in heavy industrial areas, benzo(a)pyrene in steel production areas, and fluorides in aluminum and phosphate processing areas. In addition, effects of other toxic materials, particularly those affecting the same target organ, are not considered.

3.   Finally, few if any standards take account of other physiological processes or energy demands of the job. Exposure to heat and heavier work load imposes additional environmental stress on workers. For example, workers exposed to toxic chemicals, and who are also exposed to heat so they sweat a lot, may concentrate these toxic materials to a much higher degree in the kidneys. Workers carrying out high energy functions tend to breathe more, and so take in more cubic meters of air possibly laden with pollutants than

in normal breathing. Recent studies suggest potentially
serious effects of vibration on multiple organs, and
particularly the bony skeleton. The effects of noise have
been poorly documented and assessed. Loss of hearing
represents not only a biological disease, but a loss of
social function. More effort should center on noise reduction and noise protection with surveillance of hearing loss
in high exposure situations.

3.2.4 Surveillance and Health

In addition to monitoring air and water, considerable effort must
direct itself toward surveying other potential repositories of toxic
materials, especially stable materials like lead, cadmium, asbestos,
and arsenic. Additionally, biological surveillance of vegetative,
animal, and human life must be increased in potentially affected
areas. A study of a community (El Paso, Texas) adjacent to a smelter with
apparently adequate pollution control devices, revealed, in addition
to cadmiosis, lead intoxication of a large number of children, many
of whom required hospitalization and acute treatment (Carnow et al.,
in press).   Examination of their blood resulted from a concern
about the quantity of lead emitted from the plant. Later examination
of the soil, dust, and air, revealed extraordinarily high levels of
lead.

The necessity for ongoing surveillance is related to the fact
that the most dangerous component in the human health equation may be
the chronic low-level effect. Whereas acute toxic effects alert the
community to the necessity for action, low-level effects do not.
Paradoxically, as controls become more effective, disease incubation
periods tend to increase and disease onset becomes even more subtle.
Therefore, scrutiny should continue at some high level in order to
ferret out those diseases with long incubation periods which may
possibly affect the community. Lung cancer, with a 20- to 30-year
incubation period, now kills more American males than all other cancers
combined.

Many substances are known carcinogens. These include benzo(a)pyrene,
radioactive materials, asbestos, nickel, chromates, arsenic, and
vanadium. We should intensify surveillance in the industries associated
with these materials in the light of the findings of increased
cancer in uranium, asbestos and coke oven workers. We should consider
the relation of additional contributions of carcinogens in air to these
workers, particularly since recent epidemiological studies suggest
that with each microgram of benzo(a)pyrene per thousand cubic meters
of air (an index of air pollution) a five percent increase in lung
cancer results (BEAP, 1972).

3.2.5 Coal Mine and Metal Mine Health and Safety

While the study team recognizes that health and safety measures
always involve tradeoffs, the intent of the Coal and Metal Mines Health
and Safety Acts seems to focus on miners' welfare, not on company
profits. Gross neglect of this intent in the past has caused the serious
error of ignoring the effects of coal dust on miners, resulting in
a near-epidemic of pneumoconiosis (black lung). As a result of this
neglect, some 200,000 or more miners suffer this industrial disease,
causing the nation considerable economic loss in production, payout
and additional medical costs.

The mining industry in the United States has a far higher accident
rate than other industrialized nations. And our lead-zinc and uranium
mines have as many accidents as coal mines. Furthermore, evidence
grows that production-acceleration practices increase accident rates.
Mineral processing plants also tend to have accident rates that are
significantly higher than the "all-manufacturing industry" averages.

Moreover, we question whether production cost in order to comply
with federal health and safety provisions has increased as much as
recently claimed. The ease with which most coal mining firms have com-
plied with former "unattainable dust standards" suggest that technology can usually reduce the "higher costs" of health and safety standards (USDI, 1971).   Reaching of these standards will probably
prove good business for the mining industry.

3.2.6 Health Costs

While we can place no dollar figure on human suffering and disease,
we can point out corollaries. Recent studies have begun to quantify the
health costs of air pollution (Lave and Seskin, 1970). Chronic
bronchitis and emphysema are the fastest-growing diseases in this
country. They have doubled every five years since World War II, and
account for the second highest number of disabilities under social
security.

Moreover, respiratory diseases account for more days away from
work and school than all other disease combined, and cause more than
1 billion days yearly lost from work. Studies in Chicago and elsewhere suggest a significant percentage of these illnesses relate to
industrial air pollution, particularly SO_2 in combination with particulate (Higgins, 1971). Recently a national EPA health study team
suggested that a doubling of acute respiratory infection occurred
in preschool children on high pollution days (CHESS, 1973). Relating
only ten percent of these days lost from work to air pollution shows
a cost of 100 million days of work lost. Recent studies in Chicago,
involving average air pollution figures, found that on the ten
days of highest air pollution for a given month, an average of eight
deaths per day from heart attacks occurred (Carnow and Carnow, in press).
This is a significant increase over the whole month daily average.
Since more than a half-million Americans die from heart disease each

year, many in the productive age range 40 to 65, a considerable
loss in skills occurs.

The costs for coal miners disabled from pneumoconiosis ranges
in the billions. Other areas of major concern include lead poisoning
Studies in two Illinois cities showed excessive blood lead levels in
one third of the children tested, with twenty percent requiring
treatment (Fine, et al., 1972). Based on these findings, one may
project a probable two million children with abnormally high blood
lead levels in other cities across the country.

3.2.7 Health Tradeoffs

Where no possibility of complete effluent containment exists, there
may be a choice of disposal through air or water emissions. Here,
in considering tradeoffs, a knowledge of absorption factors and organ
toxicity is valuable. For instance, lead when ingested results in
5-10 percent absorption and enters a portion of the blood circulation
system involving the liver so an additional protective block exists.
However, intake by inhalation results in absorption of from 30 to 40
percent and enters the major blood supply directly and can begin its
toxic effects immediately.

Concerning substitution for reasons of health, lead again is
a prime example. This toxic metal is ubiquitous and increasing
in urban communities. Documentation exists citing increases in blood
levels and other organs in individuals in urban centers exclusively.
Because lead moves through many systems, and tends to be cumulative,
at all levels of intake these multiple sources add to the total body
burden and increase the potential hazard. This situation requires a
serious call to action, such as the establishment of national ambient air
standards for lead and the reduction of lead in automobile fuel even if
it reduces engine efficiency.

3.2.8 Indirect Health Effects

Many of the toxic materials discussed, including metals, particulates
and noxious gases, affect plants, wildlife, and other life forms.
Nonvegetation, polluted odoriferous communities, and lowered aesthetic
values tend to diminish the quality of life. The World Health Organization defines health not only as an absence of disease but a state
of mental and physical well-being.

The most commonly studied air pollutants are ozone (O_3), sulfur
oxides (SO_2, SO_3, H_2SO_3, and H_2SO_4), and fluorides (HF, SiF_4, and water
soluble fluoride particulates). At low concentrations (e.g., 0.3 ppm
for SO_2; 0.24 ppm for O_3; and 0.5 ppb for HF) in the ambient air, these
pollutants, singly or in combination for short periods of time, can
damage the foliage of many plant species. Many studies on plant damage
from these pollutants deal with macroscopic damages; others demonstrate
that several microscopic pathological effects occur within the host
plant.

Air pollution costs United States growers a minimum of $132,000,000 annually in direct injury to plants and crops. Damage to commercial crops comes to $85.5 million, about one half of one percent of the total value of United States agricultural production. Damage to ornamental plants accounts for the remaining $46.5 million. Among commercial crops, citrus fruits are hardest hit with yearly damages of $28 million (Benedict, et al., 1971).

3.2.9 Effects of Fluorides on Vegetation and Animals

Approximately 155,000 tons of anthropogenic fluoride is emitted to the ambient air annually from steel production (41%), burning of coal (17%), phosphate plants (14%) and primary aluminum smelters(10%).

Microscopic Effects

Weinstein reviewed most of the published data on the effects of hydrogen fluoride (HF) as an inhibitor of plant enzymes, chlorophyll production, and photosynthesis (BEAP, 1971). HF also converts to organic fluoride (monofluoroacetic and monofluorocitric acids) in the leaf cells of pasture forage species and soybeans, and can also cause chromosomal aberrations (Mohammed, et al., 1966; Mohammed, 1968).

Solberg, et al., (1955) demonstrated that HF fumigation may cause cellular abnormalities such as hypertrophy and cell necrosis. Gordon's findings (1970) demonstrate that HF causes most of the thin-walled cells present in coniferous needles to enlarge (hypertrophy). Nuclei in these cells, as well as mesophyll cells, enlarge prior to death.

Macroscopic Effects

The visible macroscopic, pathological effects of fluoride fumigation on broadleaf and coniferous species show similarity to other phytotoxic gas-caused damage, as well as abiotic effects such as temperature, water stress, and nutrient deficiencies (BEAP, 1971). Only rarely does an investigator claim that a distinct macroscopic disease syndrome exists in the majority of HF-susceptible plant species, unless he locates a single fluoride pollution source in the vicinity of the damaged vegetation.

Studies show visible effects on plant species in areas of fluoride pollution by phosphate facilities or primary aluminum smelters. Treshow, et al. (1967), collected data around a phosphate plant in Idaho. Douglas fir trees close to the phosphate plant experienced significant growth reduction without visible necrosis to the needle tissues. The study indicated that excessive growth of the terminal stem could occur in ponderosa pine where fluoride levels were low but still above normal ambient air concentrations.

Studies in areas where phosphate or aluminum plants emit excessive hydrogen fluoride showed the following pathological effects on coniferous species (Gordon, 1967, 1970):

1. Excessive foliage necrosis, premature needle drop, and death of susceptible species within a three-mile radius of the fluoride source;

2. Excessive cone production in susceptible (ponderosa pine and larch) and semiresistant (Douglas fir and lodgepole) trees growing within a six-mile radius of the fluoride source.

3. Reduction of pollen viability in the semiresistant Douglas fir associated with fluoride concentration in both the cone and leaf tissues.

Many species, especially grasses, often demonstrate no visible effect even at excessively high concentrations of HF in the ambient air; however, grasses growing within two miles of a fluoride-emitting source, such as an aluminum or phosphate plant, can accumulate excessively high concentrations of fluoride (Gordon, 1970).

Fluoride Toxicity in Animals

A highly specialized field of veterinary medicine, in which approximately three dozen veterinary scientists throughout the world have carried out the majority of studies, investigates clinical and pathological effects of fluoride toxicity in animals. These scientists generally agree that forage with 30 ppm of fluoride, fed continually to cattle, horses, and sheep, damages teeth and bones, especially in younger animals.

Studies show that domestic animals raised in nonfluoride-polluted areas accumulate less than 1,000 ppm of fluoride in their bone tissues in a lifetime, but in contaminated areas, the animals' bone tissues accumulate from 1,000 to above 8,000 ppm, depending on pollution intensity.

Gordon's (1970) studies of fluoride accumulation in bone tissues of over 3,000 wild indigenous animals disclosed that:

1. Indigenous animals such as mice, rabbits, squirrels, chipmunks, deer, elk, bear, antelope, coyotes, and grouse collected where vegetation contains 10 ppm or less fluoride averaged less than 300 ppm fluoride in their bone tissues.

2. Similar indigenous animals, living in a fluoride-polluted area, accumulated two to five times more fluoride in their bone tissues than local domestic animals.

3. Deer and snowshoe hares eating forage with a 30 ppm fluoride
    concentration accumulated 2000 to 5000 ppm of fluoride in
    their bone tissues, and grouse (blue and ruffed) feeding
    on similar contaminated forage accumulated 3000 to 7000 ppm
    in their bone tissues.

3.2.10 Effects of Sulfur Oxides on Vegetation

The United States Department of Health, Education and Welfare prepared
the most comprehensive review of sulfur oxides and their effects on
vegetation, man, and animals (APCA, 1969). Each year approximately
30 million tons of sulfur dioxide enter the United States' atmosphere.
Approximately 52 percent of this SO_2 comes from coal burning, electric
power generation, and space heating. Twenty percent comes from
petroleum refining and use, and about 13 percent (or 4 million tons)
from nonferrous smelters.

The area surrounding the copper-nickel smelters of the Sudbury
district, Ontario, Canada, has probably received the most study of any
SO_2-polluted area. According to Linzon (1972), 2.7 million tons of SO_2
enters the Sudbury environment each year, damaging some conifer species,
particularly white pine, as far as 25 miles from the SO_2 emitting
sources. Linzon divided the Sudbury area into three SO_2 fume zones.
In the inner zone, of 720 square miles, white pine species had severe
SO_2 foliar damage which reduced radial and volume growth and caused
excessive tree mortality. This loss of timber yield due to SO_2
damage had a value of $117,000 per year. The intermediate zone
(1,600 square miles) had only slight SO_2-caused plant damage, and the
outer zone no SO_2 plant damage.

Hedgecock (1914) demonstrated that an SO_2-emitting source completely
destroyed 7,000 acres of nearby deciduous trees and on 17,000 adjacent
acres grassland species replaced the native forests over a period of
several years. A study by Gordon (1972b) in eastern West Virginia,
site of several coal-burning electric generating plants, concluded
that while SO_2 fumigation caused severe to moderate necrosis of the
foliage of several conifers and broadleaf species, sulfuric acid
rains caused the major plant destruction in a 300 square mile area,
extensively damaging tens of thousands of acres of commercially grown
Christmas trees. Ruston (1914) demonstrated that sulfur acid aerosol
increased soil acidity and reduced the number of soil bacteria in
the fallout area. Gorham and Gordon (1963) also discussed the impact
of atmospheric SO_2 on soil pH and the subsequent pH effects on plant
survival and distribution. Several investigators in Norway and Sweden
(Swedish Royal Ministry for Foreign Affairs and Royal Ministry of
Agriculture, 1971), in an 18-year study on the chemistry of precipitation
throughout western Europe, have shown changes in pH values of soil
and aquatic habitats.

Gordon (1972a, 1972b, in press) in three separate studies around a lead smelter, a power plant, and a kraft mill, all sulfur-emitting sources, demonstrated that SO_2 and sulfuric acid aerosol can dwarf needles, reduce photosynthesis, create pathological abnormalities to the cells of needles (such as hypertrophy and/or hypoplasia) and cause excessive accumulation of sulfate in conifer needles. Dochinger, et al. (1965), demonstrated that air pollution, and perhaps mainly SO_2 or sulfuric acid aerosol, causes chlorotic dwarf disease of the eastern white pines.

3.2.11 Recommendations

1.   Establish guidelines and standards for greatly reducing or eliminating those residual materials that can have significant effects on health, especially toxic heavy metals, fluorides, and carcinogenic particulates. If an industry cannot comply immediately it must adhere to time-cycles standards which become consecutively more stringent.

2.   We recommend a much closer surveillance of target organs (for example, the lung in coal, uranium and asbestos miners) with continuous recordkeeping so that people leaving these industries can serve in long term studies of chronic, low-level health effects. Further, it must be mandatory to continuously monitor the health of populations in high risk areas (e.g., those areas surrounding smelters).

3.   Measurement of pollutants must consider all of the toxic by-products of an industrial plant. Just measuring lead in a lead-smelter is inadequate if SO_2 and other particulates such as arsenic, zinc, cadmium and silicates remain unmeasured. Coke production necessitates a close monitoring of benzo(a)pyrene in addition to monitoring for SO_2, acid aerosols, and particulate.

4.   To evaluate particulate data, we must measure size distribution, since the degree of lung penetration relates inversely to size: the smallest particles (five micron and smaller) enter tiny air passages where they affect the lungs and seriously compromise air exchange. The smallest particles and their attached toxic materials may also reach the air sacs (alveoli) of the lungs where they are directly in contact with the circulating blood and transmitted throughout the body. In addition, the adsorbing capacity of particulate varies greatly with particle size. Ten one-micron particles have far greater adsorbing surface than one ten-micron particle, thus a greater amount of toxic substance enters the lung for the same weight of particulate.

5.   In regard to processing and fabricating, the application of known data to the evaluation of health effects of new and on-line technology becomes critical in considering tradeoffs. For example, increasing scale may increase controllability of SO_2. However, before instituting

new methods or economies of scale, we must consider the amount of
noncontainable (smallest) particulates that can escape.

6.   Limited work activity in hazardous areas must be considered for
processes using biologically cumulative toxic materials, such as cadmium, asbestos, and uranium. We need further study to quantify the
adaptability of workers as total biological entities and to identify
those susceptible to particular toxic materials because their organs
have few defenses.

     In some industries, control technology may not adequately provide
safe levels for a broad heterogeneous group of workers. Here the levels
and duration of exposure might properly gear to some quantitatively
defined scale of adaptation capability based on the capacity of
individual workers.

7.   The Mine Health and Safety Acts should require more, not less,
vigorous enforcement than heretofore. Additional mine health and
safety problems that should receive further consideration for legislation include:

     a.   the size and content of particulates in the work place;

     b.   vibration effects;

     c.   noise;

     d.   heat and humidity.

8.   If a study of the impact of high production rates, including
those stimulated by contract (piece-work) mining, shows, as we suspect,
a correlation with poor safety records, modification of such accelerated
production practices should be encouraged.

3.3   Technology Assessment and Best Available Technology

3.3.1   Technology Assessment

The study team concludes that at every stage of the materials cycle,
technology assessment should become a primary control of environmental
degradation. Current technological developments for supplying materials
to a growing economy show a lack of assessment and restraint that
can result in enormous consequences. Thus, environmental prohibitions
establish the need for technological evaluation of old and new ways
of handling mineral resources.

     Because of the changing qualitative and quantitative terms
under which industry seeks, mines, and processes materials, we

must assess technology. The growing size of mining and processing
plants and the use of several processes at one site produces ever
increasing waste material which loads our environment and leaves its
impact on the health and welfare of the entire ecosystem.

Exploration and development of such areas as the seabed,
suggest pronounced qualitative differences in possible environmental
effects. These effects have not been assessed, although developmental
technology for removing valuable material from the ocean floor has
progressed to a rather sophisticated level. Inasmuch as seabed
mining problems are of both domestic and international scope, in particular the potential world-wide effects on the ocean ecosystem, a
pooling of international expertise could perhaps best furnish strong
environmental control and enforcement. A small scale, pilot mining
system must be evaluated before any attempt at large scale recovery
is undertaken.

Similarly, on land, the use of large exploration blocks to prove
low grade reserves and the possibility of the use of nonselective mining
techniques reinforce the need for evaluation of environmental impacts
early in mine planning.

In processing minerals, industry interest will concentrate
on recovery efficiency, both as a conservation and economic necessity;
interest must also concentrate on the form and nature of the residual
materials. For one thing, industry must reduce or eliminate carcinogenic residuals or exchange them for inert materials. In addition,
nuclear mining and metallic solution mining, in situ, which differ
from older techniques especially in their nonselectivity, need evaluation of their residual effects: a monitoring on the kind and dispersion of radioactive materials and ground water contamination.

In processing, primary importance lies in the form and nature
of the residuals. For example, the particulate problems associated
with the conversion from open hearth to basic oxygen furnaces (BOF) might
have been reduced if someone had considered the assessment of the total
BOF environmental impact during the early development stage. More
dramatically, the continuing problem of SO_2 from nonferrous smelters
and coal-fired power plants necessitates a program of research and
development in both the recovery and uses of sulfur. Such a program,
which has only interested industry intermittently in the last 60 years,
must find the economic and technical means for marketing a sulfur
by-product or for storage of sulfur in a stable form, soon.

Technology assessment for environmental impact must include the
in-plant health and safety of the worker when process change is planned
rather than come as an afterthought. Product valuation must consider
the beneficial effects in health and safety and must evolve technologies and operational practices for a safer and more healthful surface
and underground working environment.

3.3.2 Best Available Technology

Mining and processing always produce residuals and these materials
impact the environment. Harmful pollutants should be monitored and
their impact alleviated. We should insist on vast improvements in
operational requirements to reduce residuals in the mining, processing,
fabrication, distribution, and final use of mineral resources to improve
the health, safety and welfare of humans. Certainly, all new plants
should use Best Available Technology (BAT) as an unalterable guideline
in the planning and installation of control equipment for toxic or
potentially harmful effluents; some mitigating circumstances over the
short-term become permissible in operating plants. We define "Best
Available Technology" as that on-the-shelf technology which most
effectively contains the emission of given residuals, or has been
demonstrated at scales sufficiently close to plant size operation so
that exprapolations fall within reasonable engineering confidence.

Another element in defining BAT, the marginal benefit on pollution
control investment, becomes an ever-present problem in the recovery
of pollutants. New technology creates a main route around diminishing
returns. Therefore, a focus on Best Available Technology encourages
a viewpoint toward economy of operation.

3.3.3 Substitution and Recycling

Components of technology assessment for environmental impact include
substitution and recycling. The study team recognizes different
problems in the recycling of mineral resource materials and the substitution of materials. Normal market conditions generally do not
consider recycling, so development of special policy considerations
could enhance recycling activities. In contrast, substitution becomes
part of the normal market options and necessitates public policy when
substitution of a material is based upon an overriding noncommercial
rationale.

We recognize that the substitution of one mineral resource for
another, in the private and public consumer market, involves considering
many factors. Despite the complexity of the problem, an urgent need
exists to include the assessments of alternative materials in the
overall planning of industry and government. For some specific uses,
such as phosphate and potash in mineral fertilizers, no substitution
is possible since these elements are required for plant growth. For
such materials, appropriate conservation methods need to be developed
to insure continuing availability over a very long time span.
Requirements for these materials have increased significantly and may
be expected to expand with, or faster than, the growth in population.

Factors we can consider in the substitution of materials include
the product's strength and chemical and physical characteristics, its
availability, cost, existing trade policies, defense priorities and
consumer acceptance. These factors are important and historically

have been considered to some extent when plans were made for the production of consumer goods and construction materials. Making comparative analyses of the environmental impacts of substituting materials has recently become more prevalent than in the past.

We must carefully consider the environmental impacts not only of substituting one mineral resource for another, but of all substitution. Such consideration must include the assessment of the environmental aspects of the extraction, processing, fabrication of products, ease of recycling and the difficulties of disposing of all materials potentially available for use in a given product. It also must include the potential redesign of a product (or process) to make possible the use of a more environmentally desirable resource.

The difficulty in making a comprehensive assessment for many materials lies in the limited knowledge of pollutant effects upon human health and the environment in general. To consider everything involved and to weigh their relative effects becomes complex and costly. Yet, the consequences of continuing to use a particular material when a more environmentally acceptable one is available, become more serious with time.

Substituting aluminum for steel in the production of containers has increased. Yet, it requires approximately twice as much electrical energy to produce an aluminum container as a similar steel container. Obviously, a significant increase in the demand for electrical energy exists because of such substitution. This in turn results in an increase in pollution associated with generating electrical energy.

In assessing the desirability of such substitution, however, we must consider other things. These include answering questions such as (1) what is the overall energy demand for each of the potential products and what are the comparative environmental impacts involved in supplying this energy? (2) which container is more readily recyclable--and more probably will be recycled? (3) what are the comparative environmental impacts of the pollutants emitted in each step of the production and fabrication cycles for each of the alternative raw materials? (4) if the commodity is not recycled, which material will have the least potential effect upon human health and the environment during its more-or-less conventional disposal?

We must answer similar questions in the substitution of plastics for nonmetallics in construction products. We should evaluate substitution not only on the bases of cost and simplicity of production alone, but also upon the environmental aspects of mining and processing of the nonmetals versus those of extracting petroleum and of petrochemicals processing techniques. The evaluation also must include problems arising from disposal and/or recycling of the discarded products before acceptance of such substitution.

A good example of the nonmarketplace role of public policy directed toward substitution is health effects due to the hazards of lead from gasoline. The Government should develop a policy to promote substitution for, and elimination of this toxic material, even though combustion engines may become less efficient. It should make similarly concentrated efforts to encourage substitute materials, or process routes, whenever a substance is hazardous to human health or impacts materially on human welfare, such as in the shifting of chlorine production in electrolytic cells from systems that allowed loss of the mercury in liquid waste streams to closed systems which recycle mercury. Industry should research and thoroughly assess materials and processes to ensure that substitutes are indeed less toxic than the materials or processes replaced. For instance, questions raised about the toxicity of NTA (nitrilotriacetic acid), the first replacement for phosphates, delayed its introduction.

With respect to recycling, pronounced institutional factors run counter to economic development in United States markets. The market problem applies specifically to obsolete scrap as opposed to prompt industrial scrap which is already adequately factored into total raw material use. Problems relating to transportation freight rates, depletion allowances for primary production, processor prejudice, consumer prejudice, and local government reluctance to implement pickup and delivery systems presently mitigate against a workable recycling industry composed of obsolete scrap. We recognize that developing an economically resilient scrap industry will affect the production and price of primary materials substantially. Viewed from environmental effects and conservation practices, institutional changes must allow obsolete scrap to become part of the normal mineral resource industry.

The study team favors increased recycling realizing that the use of such material imposes its own environmental burden. Decisions to use recycled rather than primary materials must be made on a case-by-case basis. However, system changes to promote recyclable wastes, and the growing willingness of consumers to segregate recyclable materials, will help reduce the apparent costs involved. Moreover, the variety of institutional factors noted above combine to make recycled materials apparently more expensive.

3.3.4 Solid Waste

The major solid waste burden in the materials economy occurs at the ends of the extraction-consumer chain. Large solid waste accumulations occur in the early processing stage of mineral raw materials and at the end of the consumer cycle. Those handling solid waste burdens should consider one of three alternatives. First, extract the valuable materials that the waste contains. In municipal waste, this can significantly reduce the volume of material handled. For mineral waste piles additional extraction will only slightly reduce the volume of material although it may significantly reduce the dispersion of toxic metals to the environment.

Second, find uses for the bulk wastes. Because firms try to avoid
the costs of conventional disposal, there is a market incentive to promote the use of fly ash, smelter slag and mine tailings. These attempts
have had varying results; e.g., blast furnace slag is now wanted as a
road base. Nevertheless, a need continues for research in the use of
other wastes.

Third, dispose of the solid waste so as to make recovery feasible
at a later date. This may entail sorting prior to burial. Such primary sorting may add little to the cost of disposal, but may greatly
help future ability to use the stored mineral resource. Some mining
and metallurgical firms already operate in this way, but the practice
should be greatly expanded.

3.3.5 Manpower

This section focuses on the need to train materials resource engineers
and geologists to consider the environmental aspects of their work.
Current engineering and earth sciences curricula, already full and rigorously structured, give the student little time to consider the economic, ecological or social implications of his related study fields
and the professional work he may later engage. Unfortunately, a common
disdain for "softer" environmental or social science information among
engineering and earth science students and faculty is fed by the mis-
conception that health and safety, and environmental concern prevent
engineering optimization. This attitude does a disservice to the engineering profession, for an engineer responds to human needs or demands
when he maximizes environmental protection just as when he maximizes
production.

The philosophy of "quick and dirty" has generally pervaded educational institutions that have trained mineral resource engineers for the
past 100 years. In other words, get the material out of the ground and
to market cheaply and rapidly. We cannot emphasize too strongly that
this attitude is no longer acceptable and may account in part for some
of the decrease in enrollments in mineral engineering.

3.3.6 Recommendations

With respect to technology assessment and the application of best available technology we recommend the following:

> We must assess the total environmental impact of each material
> cycle in a quantitative manner to ensure the proper environ-
> mental consideration in the introduction of large scale, new
> technologies in mining and processing of material resources.

2. Specifications for material supplies by government agencies
    should encourage markets for products that reduce all forms
    of residuals in the material resources field. These agencies
    should give paramount considerations to the general health
    and welfare effects produced by the residual impacts.

3. Federal funding of research and development for changing
    technologies associated with minerals production must
    include environmental assessment which focuses on through-
    put control to improve the in-plant health and safety of
    the worker and reduce the toxicity and quantity of waste.

4. In every new plant installation, and in older plants with
    few exceptions, federal guidelines for Best Available
    Technology for pollution control must be applied when toxic
    or potentially harmful emissions result from operation.

5. We advocate the formulation of a federal policy for reducing
    nonrecyclable convenience packaging composed of both re-
    newable and mineral resource materials.

6. We recommend studies on the feasibility for the redesign
    of products and their containers to encourage recyclability
    through improving the quality of the material and/or
    reducing recycling cost (e.g., localization of copper in
    automobile production permits easier removal and enhances
    the recyclability of the automobile; the standardization
    of liquid containers facilitates recycling and improves
    container quality).

7. We recommend a federal program to guide solid waste disposal
    in a way facilitating recovery of potentially valuable
    material at a later time.

8. We recommend a program of educational support and manpower
    retraining in order to supply the market with qualified
    technicians and professionals to fill the critical shortage
    of manpower needed in the field of recycling and substitution.

9. Every engineering curriculum should contain a specialized
    course in the ecological and social impacts of engineering
    practices and engineering works. This course should
    probably include systems training in technology assessment,
    and the Engineer's Council on Professional Development should
    require it for the accrediting of an institution's engineering
    program.

10. We also recommend the establishment of interdisciplinary
    training at the graduate level in the broad field of material
    resources and environment by institutions around the country
    with support from the federal government.

3.4 Land Use Planning

The problem of land use planning with respect to mineral resources
entails the following four important elements:

1. Land tenure and associated rights;

2. Inventory and protected reservations of
   mineral supplies;

3. Conflicts in siting;

4. Operational restrictions as a function
   of site, time and reclamation.

3.4.1 Environment, The Right to Mine, and Land Tenure

The identification of in-place mineral resources does not necessarily
mean we have determined the highest possible use for the land. No
right exists to mine on the assumed basis that mining achieves highest
economic use. The Mining Law of 1872 and the Leasing Act of 1920,
which together essentially form mineral tenure recognition for mining
enterprises on federal land, are conceptually and operationally outmoded.
Both encourage economic inefficiencies and unethical speculation with
respect to development of mineral resources. Operation under the two
laws causes damage to the environment as well as lost revenues to the
public. There is a lack of both legitimate control and public guidance
in land use on the public domain.

The study team concluded that alienation of public domain into
private ownership through the mining law should come to an end. We did
not review all possible land tenure arrangements applicable in the
United States; however, a well-formulated location system (Canadian
type), and a better leasing system than the Leasing Act of 1920, in
combination with federal and state land management responsibilities,
should be considered. Draft legislation to revise the mining laws is
pending in Congress.

In the interest of conservation and reduction of continuing environmental impact, contractual leasing obligations should utilize potential
multiple resources of each mining site. Many strip mining operations
remove the most profitable beds leaving materials behind, which, considering a slightly lower average return and greater conservation values,
could be mined. High grading causes great economic and environmental
resource waste and should be eliminated in the first cycle of mining in
both surface and underground operations.

3.4.2 Recommendations

1.   Congress and the Executive should implement new arrangements for
mineral resource ownership and management through a leasing system on
the public domain. Leasing arrangements should make provisions for
greater protection of environmental amenities in the contractural
obligations of the parties. The new law should also include adequate
bonding and policing provisions to ensure enforcement of the contracts.
Public comment on the mining and reclamation plans must precede final
contractual agreements.

2.   We recommend that any leasing or other land tenure arrangement
encourage the conservation of mineral resources through mining practices
that remove all possible valuable materials during the first mining
cycle. Freedom for society to decide when to pay the added costs of
such conservation requires disclosure of information on the nature and
extent of the potential resources with sufficient protection to ensure
the rights of the proprietor.

3.   Any new law should recognize selective withdrawal of certain sites
from mining as necessary and desirable when control of serious environmental impacts appears unlikely or higher social values for the land
exist.

3.4.3 Inventory and Protected Reserves

Many well-known cases show the removal of valuable mineral resources
from possible development through purchase by private owners for other
land use or through oversight due to lack of inventory. Clearly, the
future economy has lost valuable mineral resources, particularly the
large bulk, low unit value types necessary for construction in urban and
suburban centers. Lack of integrated land use planning will see more
of these resources lost from possible productive service.

Several local governments have created reservations of sand and
gravel (Flawn, 1970). No similar systematic approach to this problem
for other kinds of mineral resources exists on a national or regional
basis. Implementation of mineral reservations requires a high quality
inventory of the resource to allow for long-term planning. Such an
inventory would not be exhaustive, but would concentrate on areas that
lie on the fringe of rapidly expanding urban and suburban regions today
and those that seem likely candidates for future population expansion.

This principle applies to more than just sand and gravel reserves.
For example, the absence of such planning has created enormous problems
in Pennsylvania and elsewhere, where abondoned underground coal mines
lead to extensive subsidence and destruction of valuable surface
property. In such cases adequate land use planning could prevent
construction on top of old or working mines until the area is stabilized.
Similarly, in some areas where recreation is the main use of a surface

without unique natural features, society could adjust the boundaries of
the recreation areas to allow for mineral development if adequate alternative recreation sites exist.

The discussion on inventory has presupposed two problems. The first
concerns environmental controls appropriate during the exploration stage
of the materials cycle. The second concerns the creation of an information bank on mineral reserves and resources which will increase the
efficiency of mineral exploration and development and minimize the environmental impacts incurred in exploration; e.g., the French BRGM
(Agard and Vannier, 1969).

The need for an information bank is evident; rational land use
planning cannot proceed without resource information. However, we
concern ourselves here principally with the fact that the adverse environmental effects of exploration are often magnified because of repetitive
examination of the same land by commonly destructive techniques. The
additive effect of such impacts may be of the same order of magnitude
as the concentrated environmental impact in a single location resulting
from mining itself. Therefore, we must save and synthesize pieces of
information about mineral resources, while at the same time protecting
the confidentiality of this information for a reasonable length of
time. Creating exclusive exploration rights in certain areas, with
data filing requirements, or requiring the disposition of a certain
portion of the drill cuttings or cores in a government facility are
two possible approaches.

We should encourage exploration techniques that lessen environmental impacts. Developing exploration methods, such as remote
sensing, geochemical and geophysical surveys, use of helicopters and
of backhoeing rather than bulldozing, lessen the adverse effects per
unit of information recorded. Nevertheless, drilling is eventually an
integral part of exploration. Another way in which environmental effects
of exploration come to be magnified is through a "staking rush." Team
members believe that the inordinate haste in which exploration is often
conducted leads to a variety of short cuts that cause severe environmental problems. Moreover, there are some particularly fragile areas,
notably polar and alpine tundra regions, estuaries, offshore areas, and
areas of high water quality, where special restrictions must be applied in
order to protect the environment. Throughout this discussion there is
an underlying assumption: one accepts the environmental impact of
mining operations because of the relatively predictable return for other
sectors of the economy; however, in exploration there may be no
return, therefore the impact on the environment should be minimal.

3.4.4 Recommendations

1.   We strongly recommend the federal development of a central or
regional land planning information bank to contain mineral resource
information of similar character to that already deposited by the
petroleum industry in many states. The requirement for depositing information in the land information banks should be obligatory.

2.   With respect to exploration technology, the contractual agreements
leading to exploration rights should include the prescription for the
use of low impact technology. Legislation, with respect to exploration,
should promote environmentally economizing practices. Relatively long
time periods should be incorporated into both the exploration and
development phases.

3.   The federal government should institute a system of mineral reservations which regulates interim uses of the land in a manner
compatible with sequential or later development of the resources. Conflicting uses such as the construction of permanent facilities over
the mineral deposits should be discouraged.

3.4.5 Conflicts in Siting

The reading of recent newspapers make us recognize that the siting of
industrial plants, particularly power plants, concerns many people.
The study team could identify three general circumstances for prohibiting
mining or the use of certain mining techniques or equipment:

    1.   Where mining conflicts with a "higher" or "better" use of
       land. Examples of this conflict include wilderness areas,
       wetlands, certain coastal terrains and terrains containing
       special water quality systems (e.g., scenic and wild rivers).

    2.   Where it is all but impossible to contain the impacts that
       would result from mining activities; examples include
       estuarian areas, steep slopes, and certain arctic environ-
       ments.

    3.   Where conditions make it unlikely that land stabilization
       and reclamation will occur within reasonable time and
       cost frameworks (for a definition of reclamation see
       below). Examples include surface mining on steep slopes,
       some steep walled rock quarries near urban areas, and
       surface mining in unusually fragile environments.

In these cases, mining will not best utilize the land. In some of these
circumstances (as with rock quarries) removal will occur on a case by
case basis. For others, such as most wilderness areas, the only operational possibility will be to exclude large blocks of land at one time.

We will stress a further point which strengthens our conclusion:
the value to society of its remaining special land areas will increase
as population increases and as the supply of such land, which we find
in fixed quantity only, decreases (Krutilla, et al., 1972). In addition,
we can reverse almost any decision to preclude mining, but in most
cases we cannot reverse the effects of a decision to alter the prevailing
state of the special land areas.

A closely related problem arises with the siting of both resource-
based communities and processing plants. A variety of problems need
consideration, including total environmental impacts created by the
service population and infrastructure that the facility will require for
its operation as well as the type of processing planned.

Plant and mine density needs consideration in each region. Depending
on the likely effluents and the size and efficiency of potential treatment
facilities, we might consider clustering the polluting plants rather than
dispersing them. For highly impacted areas, we should consider integrated
treatment plants and unit marketing of waste products.

A further environmental zoning technique, the creation of protective
zones around a plant, deserves consideration particularly where it only
indirectly concerns human health. Such areas might preclude the planting
of certain kinds of vegetation or the grazing of certain animals or even
human habitation itself, depending on their comparable sensitivities.

Additional planning considerations must enter into the decision to
locate resource and extraction communities to insure social, economic and
political continuity. By communities we mean those settlements with
families and intended permanence of location. A community should not
start unless a reasonable expectation exists that continuity can be
established and can lead to a diverse economic base. For example, the
following kinds of diversity are possible in such a community with a
skilled labor force: (1) capital intensive light industry, (2) health
and educational services, and (3) a mixture of tourism. In terms of the
above criteria, the community structures related to coal strip mining
sites in the western United States presents obvious problems. This
does not preclude mining in isolated areas; there are alternatives
such as planned temporary communities or air bussing of the labor force.

3.4.6 Recommendations

1.   Land use planning should always allow that at a given point in time
and geographic-geologic-cultural circumstances, mining is not an
appropriate land use. In this situation explicit withdrawal criteria
should exist and the lands withdrawn from mining should be clearly
marked and publicized.

2.   Governments should base the siting of extractive processing facilities
on a comprehensive land use planning decision in which ecological and societal,

as well as market considerations play a role. Baseline ecological
studies should be performed on alternative sites prior to plant location
selection in order to make comparisons between preoperational and
operational situations, and to use such information in the final plant
site selection procedure.

3.   In the planning and development of new production facilities with
possible toxic emissions, geomorphic and meteorological factors of the
impact area require consideration (e.g., wind direction, wind speed,
inversion frequency and depth of the mixing layer, slope of ground
water table).

4.   New resource-based communities should not be created unless planning
results in a reasonable expectation for their long-term viability.

3.4.7 Operational and Reclamation Requirements

Reclamation must become part of the process of mining and planned into
the mining system. Mining restrictions should depend upon the geology of
the mineral deposit and local conditions of geography, climate, soils, and
population density. Bedded deposits which extend horizontally in two
dimensions often impose a greater environmental burden, whether surface
or deep mined, than do more equidimensional deposits. Generally, we
draw no distinction in what follows between surface and underground
mining. The specific effects and control techniques may differ but the
objectives of environmental management remain unchanged.

   For the purpose of our discussions we have developed the following
definition of reclamation:

> Restoration of the land, at minimum, to a stable and
> diverse ecological unit, no less productive (econo-
> mically and ecologically) than before disruption.

We recognize that in some cases the economic productivity of land
approaches zero; in no case does the ecological productivity drop that
low. In many cases reclamation of land to a higher value might prove
wise or even profitable. (cf. chapter 4.2.3)

   In the past, industry has surface mined large parts of a given
region of the United States and left it unreclaimed (estimated to be over
3×10^6 acres, Senate Subcommittee Hearings, 1971). The federal government surveyed these areas and prepared preliminary cost estimates for
reclamation to minimum standards. Generally people agree that only the
federal government holds enough funds to accomplish the reclamation of
orphan areas. However, another opportunity exists for reclamation in
active mining areas. When a new mining firm moves into an area already
mined and left inadequately reclaimed, the government conceivably can
induce the firm to undertake complete reclamation activities as a part
of its ongoing operations; for example, to undertake water quality
improvements by treatment of area acid mine drainage in a single treatment
plant or the grading and planting of old tailings or spoil banks.

Large volume, low value waste products that typically result from mining and processing operations create special problems. Examples include: concentrator tailings, smelter slags, phosphate slimes, red muds, and others. In all cases, these materials require handling to minimize the off-site effects. Well developed techniques exist to stabilize and control tailings, for instance replacing wastes in a void. If disposal of tailings results in acid drainage, fire, or particulate contribution to water or air contamination, it probably shows mishandling (Bureau of Mines, 1971). Certain kinds of wastes, notably uranium tailings, require special precautions.

3.4.8 Recommendations

1. Federal law should require that handling of waste materials from mining and processing operations be done in such a way that there are essentially no continuing off-site effects.

2. The federal government should inaugurate a program to provide fair recompense for mining firms that undertake environmental quality improvements where the land has been previously mismanaged.

3.5 Economic Considerations

The study team is aware that all of the foregoing has underlying economic repercussions. This section, rather than surveying these effects (see chapters 2 and 6), highlights a few points that need consideration.

The first general conclusion is that the demand for, and thus the value of, environmental enhancement steadily increases with time. Mineral materials are only one of the goods and services produced in a market economy and there exists no a priori reasons to value them more than other goods and services, human longevity or other environmental resources. Obviously, over the past decade, government increasingly responded to the increased value placed on environmental resources relative to material resources through many actions in the environmental, public health and welfare fields.

Whatever the economic alternatives related to environmental enhancement, they should reflect a system that concentrates on inputs and throughputs rather than only "end-of-pipe" containment. It will be more efficient in economic terms, and good environmental management, to focus on process material and the processing method rather than the residuals from a given system.

An environmental management system that sets controls on residuals should contain two principal elements: (1) a federally backed system of industry-wide minimum standards to ensure that no pollution havens capable of inflicting widespread damage or exacerbating health problems exist; (2) a system to allow state or local governments to institute additional regulations on siting, management, or emission controls.

The cost of pollution control in the minerals industry remains within a range absorbable or acceptable to consumers. Today technology already exists to greatly ameliorate our present environmental problems and, if adopted by industry, would not endanger its economic viability. Moreover, recent research on the nature of large corporations suggests that certainty, rather than cost itself, determines success. A pollution control measure then should create an atmosphere in which industry can expect that a certain requirement, standard, or tax will endure for some time.

Beyond the minimum considerations described above, the right to choice remains paramount. An individual chooses both living and working conditions that imply tradeoffs for himself. For example, he might choose a more dangerous occupation in return for a higher salary. However, in encouraging freedom of choice above the minimum standards, we must ensure a truly free choice. Therefore, at least three conditions must be met:

1. The choice must be made with full information.

2. The choice must not be so constrained that the person really has no alternative, as for example, between working and no income at all.

3. One has some freedom from the tyranny of the many small choices made by others.

References

Agard, J. and M. Vannier, 1969. *Instruction pour l'inventaire des gites mineraux de la France*. (Orleans, France: BRGM Publ. No. 69 SGL 182, GIT Bureau de Recherches Geologiques et Minieres), 32 pp.

Air Pollution Control Administration (APCA), 1969. *Air Quality Criteria for Sulfur Oxides*. (Durham Research Triangle Park: Publication AP-50, APCA), 178 pp.

Benedict, H. M., R. E. Olson, and C. J. Miller, 1971. *Economic Impact of Air Pollutants on Plants in the United States*. Prepared by Stanford Research Institute for Coordinating Research Council, Inc. (Washington, D. C.: NTIS PB 209 265), 81 pp.

Carnow, B. W. and V. A. Carnow, in press. "Air Pollution Morbidity and the Concept of No Threshold," *Advances in Environmental Science and Technology*, v. 3, Pitts and Metcalf, eds. (New York: John Wiley & Sons, Inc.)

Carnow, B. W., V. A. Carnow, and B. F. Rosenblum, in press. "Unsuspected Community Lead Intoxication and Emissions from a Smelter: the El Paso Study," *Archives of Environmental Health*.

Committee on Biologic Effects of Atmospheric Pollutants (BEAP), 1971. "Effects of Fluoride on Vegetation," *Fluorides*. (Washington, D. C.: National Academy of Sciences), pp. 77-132

Committee on Biologic Effects of Atmospheric Pollutants (BEAP), 1972. *Particulate Polycyclic Organic Matter*. (Washington, D. C.: National Academy of Sciences), 361 pp.

Community Health and Environmental Surveillance System (CHESS), in press. *Health Consequences of Sulfur Oxides: a report from the CHESS Program 1970-1971*. Prepared by the Human Studies Laboratory, National Environmental Research Center, Environmental Protection Agency. (Research Triangle Park, North Carolina: Environmental Protection Agency)

Council on Environmental Quality (CEQ), 1972. *Environmental Quality: the third annual report of the Council on Environmental Quality*. (Washington, D. C.: U. S. Government Printing Office), 450 pp.

Dochinger, L. S., C. E. Selishar, and F. W. Bender, 1965. "Etiology of Chlorotic Dwarf of Eastern White Pine," *Phytopathology*, v. 55, p. 1055.

Fine, P. R., C. W. Thomas, et al., 1972. "Reduction in Blood Lead Levels: a study in fourteen Illinois cities of intermediate population," Journal of the American Medical Association, v. 221, n. 13, pp. 1475-1479

Flawn, P. T., 1970. Environmental Geology: Conservation, Land-Use Planning, and Resource Management. (New York: Harper and Row), 313 pp.

Gordon, C. C., 1967. Report on the Effects of Hydrogen Fluoride on Vegetation in Garrison, Montana. Prepared for an intrastate air pollution abatement conference, Powell County, sponsored by APCA. (Washington, D. C.: Air Pollution Control Administration) 26 pp.

Gordon, C. C., 1970. The Effects of Hydrogen Fluoride Pollution on the Flora and Fauna of Glacier National Park. (Washington, D. C.: Environmental Protection Agency), 411 pp.

Gordon, C. C., 1972a. "Effects of Air Pollution on Indigenous Animals and Vegetation," Helena Valley, Montana, Area Environmental Pollution Study, EPA Publication AP-91 (Washington, D. C.: U. S. Government Printing Office), pp. 95-112

Gordon, C. C., 1972b. A Study of Air Pollution and Its Effect on Flora in the Mountain Storm Area of West Virginia and Maryland. (Washington, D. C.: Environmental Protection Agency), 84 pp.

Gordon, C. C., in press. "The Biological Effects of a Kraft Pulp and Paper Mill's Emissions on Vegetation of the Missoula Valley (Montana)," Environment.

Gorham, E. and A. G. Gordon, 1963. "Some Effects of Smelter Pollution upon Aquatic Vegetation near Sudbury, Ontario," Canadian Journal of Botany, v. 41, pp. 371-379

Hedgecock, G. G., 1914. "Injuries by Smelter Smoke in Southeastern Tennessee," Washington Academy of Sciences Journal, v. 4, pp. 70-71

Higgins, I. T. T., 1971. "Effects of Sulfur Oxides and Particulates on Health," Archives of Environmental Health, v. 22, pp. 584-590

Krutilla, J. V., C. J. Cichetti, A. M. Freeman III, and C. S. Russell, 1972. "Observations on the Economics of Irreplaceable Assets," Environmental Quality Analysis, Kneese, A. V. and B. Bower, eds. (Baltimore: Johns Hopkins Press), pp. 69-112

Lave, L. B. and E. P. Seskin, 1970. "Air Pollution and Human Health," Science, v. 169, pp. 723-733

Lepper, M. H., N. Shioura, B. W. Carnow, S. Andelman, and L. Lehrer, 1969. "Respiratory Disease in an Urban Environment," Industrial Medicine and Surgery, v. 38, n. 4, pp. 126-131

Linzon, S. N., 1972. "Effects of Sulfur Oxides on Vegetation," Canadian Forestry Chronicle, v. 48, n. 4, pp. 1-5

Mohammed, A. H., 1968. "Cytogenic Effects of Hydrogen Fluoride Treatment in Tomato Plants," Journal of the Air Pollution Control Association, v. 18, pp. 395-398

Mohammed, A. H., J. D. Smith, and H. G. Applegate, 1966. "Cytological Effects of Hydrogen Fluoride on Tomato Chromosomes," Canadian Journal of Genetics and Cytology, v. 8, pp. 575-583

Ruston, A. G., 1914. "The Plant as an Index of Smoke Pollution," Annals of Applied Biology, v. 7, pp. 392-402

Senate Subcommittee on Minerals, Materials and Fuels of the Committee on Interior and Insular Affairs, 1971. Hearings on Pending Surface Mining Legislation, Parts 1-3, 1173 pp.

Solberg, R. A., D. F. Adams, and H. A. Ferchau, 1955. "Some Effects of Hydrogen Fluoride on the Internal Structure of Pinus Ponderosa Needles," Proceedings of the Third National Air Pollution Symposium, Division of Industrial Research, Washington State Institute of Technology, Reprint 35. (Pullman, Washington: Washington State University Press), pp. 164-174

Swedish Royal Ministry for Foreign Affairs and Royal Ministry of Agriculture, 1971. Air Pollution Across National Boundaries: The Impact on the Environment of Sulfur in Air and Precipitation, a report by the Swedish Preparatory Committee for the United Nations Conference on the Human Environment. (Stockholm, Sweden: United Nations Conference on the Human Environment Secretariat), 96 pp.

Treshow, M., F. K. Anderson, and F. Harner, 1967. "Responses of Douglas Fir to Elevated Atmospheric Fluorides," Forest Science, v. 13, n. 2, pp. 114-120

United Nations Conference on the Human Environment (UNCHE) 1971. Development and the Environment (the "Founex Report"), 31 pp.

U. S. Bureau of Mines, 1971. Bureau of Mines Research Programs on Recycling and Disposal of Mineral-, Metal-, and Energy-Based Solid Waste, Bureau of Mines Information Circular IC 8529. (Washington, D. C.: U. S. Department of the Interior), 53 pp.

U. S. Department of the Interior (USDI), 1971. Toward Improved Health and Safety for American Coal Miners: 1970 Annual Report of the Secretary of Interior (Washington, D. C.: U. S. Government Printing Office), 79 pp.

Winkelstein, W. Jr., S. Kantor, E. W. Davis, C. S. Manert, and W. E. Mosher, 1967. "The Relationship of Air Pollution and Economic Status: total mortality and selected respiratory systems mortality in men: I. Suspended Particulates," Archives of Environmental Health, v. 14, pp. 162-171

CHAPTER 4

STUDY TEAM

ON

ENVIRONMENTAL PROBLEMS ASSOCIATED WITH FUEL MATERIALS

C. WAYNE COOK, Department of Range Science, Colorado State University,
    Fort Collins, Chairman
RICHARD C. AUSTIN, Appalachian Strip Mining Information Service, Seth,
    West Virginia
VINCENT M. BROWN, Executive Director, National Petroleum Council,
    Washington, D.C.
JAMES R. GARVEY, Bituminous Coal Research, Inc., National Coal
    Association, Monroeville, Pennsylvania
GERALD F. GIFFORD, Watershed Science Unit, Utah State University, Logan
JACK KELLER, Environmental Conservation, Coordinator, Exxon Corporation,
    U.S.A., Washington, D.C.
CHARLES H. PRIEN, University of Denver, Colorado
RALPH A. LLEWELLYN, Staff Officer

4.1 Energy Overview

It is generally agreed that the United States is, for a combination of reasons, confronted with serious problems concerning the location, acquisition, allocation and use of nonrenewable fuel resources.

The past de facto national policy that the availability and cost of energy should not be constraints to development is now the subject of long overdue examination. Quite clearly, the historic approach will no longer work. While domestic reserves of fuel materials are large in terms of current annual use rates (Table 1), they are finite. Concern for environmental protection is significantly altering the economy's fuel mix and is posing the possibility of substantial dislocations in some segments of the energy industry. For the first time in its history, the nation finds itself depending on significant amounts of imported fuels to meet demand and is facing the prospect that this dependence will almost certainly increase. Coupled to these concerns are the complex considerations of national security and international balance of payments.

Table 1: Estimated Energy Resources and Production
(x 10^{12} Calories)

Resources	Estimated Resources (x 10^{12} Calories)	Annual Production (1970)a (x 10^{12} Calories)
Coal	10,390,000[b,e]	3840
Petroleum	663,000[b,c]	4980
Natural Gas	554,000[b,c]	6080
Shale & Rock Oil	1,468,000[b,c]	--
U_3O_8	1,300,000[f]	60
ThO_2	1,462,000[d]	--

Sources: a USDI, U.S. Energy Through the Year 2000, 1972.
b USGS, Circular 650, 1972.
c National Petroleum Council, 1972.
d USDI, Mineral Facts and Problems, 1970.
e National Coal Association, 1970.
f Atomic Energy Commission, 1970.

While some aspects of the energy problem bear directly on environmental matters, nearly all have implications for the environment. It is important to realize that, although the United States does not have a coherent energy policy, it does have an enunciated environmental policy defining the acceptable qualities for air and water and is beginning to formulate pieces, at least, of a land use policy, in particular with regard to surface mining and extraction of oil in the coastal zones. The study team concluded that the absence of a broad national energy policy is a major factor in the impending energy problem.

We recognize that the definition of such policy will be difficult. There is already in existence a variety of national policies, each adopted with consideration for prudence and expediency in gaining stated social benefits such as full employment, economic growth, and an unimpaired environment. As might be expected, there are many internal inconsistencies arising from changing values, knowledge, and technology. The establishment of a national energy policy could add to these inconsistencies, conceivably in major ways. Nonetheless, we see a clear urgent need for its prompt establishment in order to enable informed choices in the critical years ahead.

A national energy policy should be formulated in terms that are appropriate for achieving proper balance between the many national policies and for optimum resolution of conflicts between resources and needs. We, therefore, make the following recommendation:

Recommendation

We strongly urge the prompt establishment of a comprehensive and coherent national energy policy consistent with existing national policies relating to the quality of the environment, the health and welfare of the people, and the economic health of the nation.

While the National Commission on Materials Policy may certainly adopt the recommendation stated above and might well further recommend specific elements of a national energy policy, the study team did not feel that the commission realistically could be expected to have a decisive impact on the definition of that policy. Its statutory lifetime is limited whereas several ongoing agencies with vested interests in national energy matters are marshalling internal efforts aimed at giving them strong future voices in the design of national policy in this area. In addition, the lead times involved in effecting even moderate rearrangements in the overall national energy picture are very long indeed, and the commission will be in no position to monitor future energy development or to follow up its recommendations with any authoritative or persuasive ability.

For these reason, the study team chose to view its task with a certain amount of pragmatism. We individually held views regarding the desirability of a number of widely debated, future energy options. Nonetheless, we agreed unanimously that certain developments would occur over the next 10 to 12 years regardless of suggestions we might make, but that the actions taken now could have profound effect on our energy situation beyond 1985. In particular, we concluded from our analysis of a recent report by the Office of Emergency Preparedness that no energy conservation strategies that might be imposed (short of national emergency procedures) would have significant effect on the country's energy demand over the short term; i.e., through about 1985 (OEP,1972). The study team considered it absolutely essential that energy conservation programs be developed expeditiously with priority attention being given to petroleum and natural gas. Very recent projections

(National Petroleum Council, 1972) predict that energy requirements in 1985 will be 125 quadrillion Btu, an increase of nearly 85% over 1970. While we believe these projections may be too high, it does seem certain that total energy requirements will continue to rise, at least through 1985. In addition, while work is underway on a number of promising projects, we foresee no significant augmentation of existing energy sources from new technologies or previously untapped sources during the same period. We do foresee severe strains on the energy production system over the next decade.

It must be emphasized that we did not consider the possible effects of increasing costs and prices upon the "demand" for energy. This is not to say that we did not realize that the costs and prices will increase and that these will result in some short run (through 1985) reduction in per capita energy consumption. We also realize that the long run effects from increasing energy prices may have a significant influence on energy demands. However, the supportive studies on the price-sensitivity of energy consumption are not yet complete enough to give us a base from which we can make reliable statements regarding the effects of increasing energy costs.

It has historically been the practice of this nation to rely exclusively on domestic fuels to meet energy requirements. This practice has obviously been amended in recent years, since we are now importing more than 25% of our petroleum demand, a situation which is currently the subject of high level review. It was our feeling that the evolving national energy policy will continue to include a heavy reliance on indigenous fuel supplies.

We discussed at length the suggestion of not surface mining, not drilling for oil off-shore, not building new pipelines, etc. We agreed that the increasing difficulty in fulfilling the national demand for indigenous fuel materials of all kinds while meeting established land, air, and water quality standards will result in severe pressures for extraction by all available methods. In this context, surface mining presents distinctive problems due to the large areas of land that might be affected—the impact of scale. Coal surface mining is growing and is likely to grow at a particularly rapid rate in the western states. While our discussion was primarily directed at near-term problems and was tempered by the appraisal that the near-term elimination of coal surface mining is unlikely, we did recognize that the abundance of coal reserves of all ranks (see Table 5) and the diversity of environmental circumstances surrounding surface extraction make it incumbent upon the Federal Government to establish regulations which would prohibit surface mining in areas where the environmental impact is most critical (such as steep slopes, designated wildernesses, and watersheds already imbalanced) while rigorously controlling the environmental impact of surface mining in other areas. The technology to recover and burn these resources is available at some cost, so that a rational discussion on tradeoffs with environmental impacts can be made. On the basis of the foregoing we make the following recommendations. The first of these suggests the policy; the next two indicate the necessary implementing actions.

Recommendations

1. We urge that energy policies deal specifically with controlling environmental impacts, protecting those areas where the impacts of fuel extraction are unacceptable with present reclamation technology, maintaining consistency with long-term national energy policy, and minimizing economic dislocation.

2. Coincident with the above, it is essential that we prepare and implement national programs for demand modification strategies, including energy conservation, pricing policies, and better use efficiency, in order to bring demand in line with sustainable supplies.

3. Development must be coupled with close environmental monitoring directed at the control of impacts on land, air, and water quality. This is particularly important if development is aimed at maximizing the usability of domestic primary and secondary energy sources.

We must emphasize the need for caution with regard to a closely related factor, namely, effects of scale of development. The natural ecosystem has a recognized ability to absorb a certain amount of disturbance, whether it be alien chemicals, land disruption, or visual impacts. This ability is not limitless or well understood. There are indications that the environmental effects of a given action may increase faster than in direct proportion to the intensity of that action, particularly as natural assimilative capacity approaches saturation. There is also the matter of simple addition; i.e., while one electrical generating plant in a particular area may not cause local SO_2 concentrations in excess of air quality standards, two or more plants may do so. Care must be taken in developing the fuel resources that scale factors do not inadvertently cause violation of established environmental standards.

4.2 Overall Policy Guidelines

4.2.1 Decision Criteria for Long-Term Fuel Material Development

A national energy policy should have a sense of purpose in dealing with energy supplies, development, demand, efficiency of utilization, and environmental protection.

A national long-term energy policy must consider short-term goals as a working base. Such a policy should be a cooperative endeavor by government,

industry and the public. There is indeed a basic need for education
and cooperation in developing a common understanding of the social benefit
and necessity of utilizing our energy resources effectively, but consistent
with supplies and needs.

It is not desirable to continue using oil and natural gas for the
generation of electricity in view of the substantially greater reserves
of coal and nuclear fuels and the relative substitutability of the
several fuels in the energy system.

It appears that a source of energy at the most economical cost should
not be the sole determining factor, but that the long-term requirements
of various energy sources to meet optimum standards established by society
must also be an important consideration. Imports of critical energy
sources should consider conservation of national energy reserves, environmental protection and the emergency measures associated with national
security. It should be noted that costs invoked by demands for higher
standards of environmental quality may materially reduce the per capita
energy consumption.

It is assumed that in the very long-term all of our energy sources
will eventually be needed to satisfy our national requirements. It is
further assumed that this is true even though we receive comparatively
large quantities from foreign sources. It is therefore imperative that
energy reserves be developed in a judicious manner to meet very long-term
future needs as well as near and intermediate term requirements.

Policies for the development of fuel materials should recognize the
long lead times required.

4.2.2 Coordinate Energy Policies Among Government Agencies

There are many federal, state, and local agencies that have specific
responsibilities for various fuels. Their actions are often impromptu,
duplicative, or divergent. Much of the confusion and delay that now
plagues energy suppliers stems from conflicts among government agencies.
All too often, one agency may encourage an action while another agency
prohibits it. Coordination of federal energy policies in the Executive
Branch is necessary to provide focused and consistent guidance on energy
matters to ensure that the nation's vital needs are met.

4.2.3 Identify Realistic Environmental Standards

In this context, particularly in the matter of reclaiming surface mined
land, it is important that the term "rehabilitation" be clearly understood.

Rehabilitation was defined by the study team as the restoration of the land form and productivity in conformity with a prior land use plan toward a balanced ecological and social state that does not contribute substantially to environmental deterioration and is consistent with surrounding aesthetic values. (cf. chapter 3.4.7.)

There are some very substantial questions inherent in the definition. How and by whom is the "land use plan" developed? What is the local "balanced ecological and social state"? How will we know if the result is "consistent with surrounding aesthetic values"? These questions must be answered. They will require careful, individual consideration by national panels at the appropriate times.

Environmental standards and an environmental ethic must be established if society's needs for energy and a quality environment are to be satisfied. However, programs to assure environmental quality during the exploration, extraction, and consumption of energy fuels will involve large sums of capital. So, in reordering its priorities, the nation must recognize the inescapable impact of added environmental costs on supplies and prices.

Standards for a better environment must recognize the time required to effect the desired results as well as the impact of the standards themselves in stimulating compliance. They must be compatible with such other important national goals as full employment, reduction of poverty, further improvement in average living standards, assurance of energy supplies at all times for health, comfort, and national security.

Guidelines for environmental protection associated with extraction of fuel materials should allow a no development policy when circumstances dictate that a disturbance would be unrepairable or catastrophic in magnitude. On the other hand, extraction of fuel materials should consider all alternatives and all methods that would minimize perturbations of ecological systems. Rehabilitation of extraction activities should consider feasibility and reasonability consistent with tested methods and acceptable costs.

In general, it is believed that environmental impacts are in direct proportion to fuel production activity. It is common knowledge that impact of fuel material production depends to a large extent upon the scale and frequency of development. The ability of land and environment to absorb environmental impacts will not only depend upon the scale of development but also upon geomorphic features, air currents, watershed characteristics, plant and animal population structures, aesthetic values, and social components.

The fuel suppliers are capable of operating in such a way as to satisfy the requirements of society with respect to the environment. The role of government should be to ascertain the effects of pollutants and to prescribe and enforce workable standards of air, water, and land

quality. There is a great necessity to simplify requisite regulatory
approvals by city, county, and state authorities.

4.2.4 Research Needs

This study indicates that research is needed in a number of specific areas
covered by the recommendations and in a number of additional fields as
well: (a) exploration and extraction methods that minimize surface land
disturbances; (b) more efficient production and use of energy; (c) identification of soil and plant ecosystems that either will or will not
respond to artificial rehabilitation; (d) development of methods of rehabilitation on arid land and ecosystems with respect to physical
character of spoil banks, plant species to be seeded and wildlife habitat
to be restored; and, (e) determination of the time interval for full and
complete restoration for various vegetation ecosystems. The extent to
which such research is undertaken by private industry will, however,
depend on establishment of an economic and regulatory climate that will
motivate the conduct of such research and on the development of improved
environmental monitoring technology.

4.3 Coal

4.3.1 Reserves and Production

While we were specifically not concerned with making an energy study, that
being a task of the commission itself, the current and projected reserves
and production figures for fuel resources form an important component of
the basis of our recommendations. In the case of coal, our major concern
was with its extraction and the associated environmental impacts. The
environmental consequences of using coal and the mitigation of those
consequences; e.g., the removal of SO_2 from stack gases or sulfur directly
from coal via one or another of the several methods under investigation,
were deliberately not a part of this team's charge. They are however
extremely important and must be considered in formulating energy policy.

The total bituminous, sub-bituminous and lignite coal reserves of
the United States are estimated at over 1,600 billion tons (see Table 2).
Slightly over one-half are located in far western states and about one-half
of the total is sub-bituminous and lignite. As shown in Table 2, there
is a positive relationship between the types and locations of the coal
reserves; the sub-bituminous and lignite coals are predominately in the
west and the western coals are almost entirely of these types. In addition,
much of the western coal is located in thick seams close to the surface
making them suitable for recovery by surface techniques.

Table 2:   I. Coal Reserves of the United States*
                (x 10^6 short tons)

	Bituminous	Sub-Bituminous	Lignite	Total
East and Midwest	629,254	115,207	7,490	751,951
**West	94,291	313,957	440,593	848,841
Total	723,545	429,164	448,083	1,600,792

II. Strippable Coal Reserves of the United States
            (x 10^6 short tons)

	Bituminous	Sub-Bituminous	Lignite	Total
East and Midwest	12,506	none	25	12,531
**West	1,091	24,318	7,046	32,455
Total	13,597	24,318	7,071	44,986

*   Mapped and explored; 0 - 3,000 feet overburden.
** Includes Alaska, Oklahoma, Colorado, Arizona, Montana, New Mexico,
    Wyoming, California, Washington, Texas, North Dakota, South Dakota,
    and Utah.

Reference: U.S. Bureau of Mines, 1971

Not all of the coal reserves in Table 2, I, are considered to be recoverable. Recoverable coal reserves (defined as seams more than 28 inches
thick for bituminous and anthracite and more than 60 inches thick for sub-
bituminous and lignite, within 1,000 feet of the surface and assuming
50 percent recovery) are estimated to be 390 billion tons (U.S. Geological
Survey, 1972). The strippable portion is estimated by the Bureau of Mines
to be 45 billion tons, or about 11 1/2% of the total recoverable coal
reserve. The Table 2, II shows the predominately western distribution as
well as the concentration by rank (sub-bituminous and lignite coals) of
these reserves.

While about one-half of the coal resources of the United States are
located in the west, about 93 percent of current production is from coal
seams located in the midwest and east (see Table 3). About 50 percent of
current coal production is by surface methods.

Table 3: 1971 Lignite & Bituminous Production
          (x 10^3 short tons)

	Underground	Strip & Auger	Total
East and Midwest	271,228	242,881	514,109
*West	4,660	33,423	38,083
Total	275,888	276,304	552,192

*Includes Arizona, Colorado, Montana, New Mexico, North Dakota, Oklahoma, Wyoming, and Utah.

References: U.S. Bureau of Mines. 1972.
            U.S. Geological Survey, 1972.

Because western coal deposits are for the most part sub-bituminous and lignite, they have on the average one-third less heat content than the bituminous coals of the east. On the other hand, as shown in Tables 4 and 5, the sulfur content of western coals is substantially lower and such coals comprise 88% of the low sulfur (less than 0.7%) coal reserves of the United States. As Table 5 shows, there are also substantial reserves of low sulfur coal in the eastern United States. A considerable amount of this is metallurgical coal in thin seams, but significant amounts are being mined for electrical power generation.

Table 4: Sulfur Content of United States Coal According to Rank

	Percentage of total resources		
Rank	Low Sulfur 0 - 1	Medium Sulfur 1.1 - 3	High Sulfur 3
Anthracite	97.1	2.9	-
Bituminous coal	29.8	26.8	43.4
Sub-bituminous coal	99.6	.4	-
Lignite	90.7	9.3	-
All ranks	65.0	15.9	20.0

Reference: U.S. Geological Service, 1972.

TABLE 5

Estimated remaining coal reserves of the United States, by rank, sulfur content, and State, on Jan. 1, 1965

(Million short tons)

Coal rank and State	Sulfur content, percent									Total
	0.7 or less	0.8 - 1.0	1.1 - 1.5	1.6 - 2.0	2.1 - 2.5	2.6 - 3.0	3.1 - 3.5	3.6 - 4.0	Over 4.0	
Bituminous coal:										
Alabama	889.2	1,189.3	5,421.7	5,182.8	458.8	417.4	-	-	18.6	13,577.8
Alaska	20,287.4	1,100.0	-	-	-	-	-	-	-	21,387.4
Arkansas	-	-	1,128.4	293.1	154.0	-	40.3	-	-	1,615.8
Colorado	25,178.3	37,237.2	-	-	-	-	-	-	-	62,415.5
Georgia	-	76.0	-	-	-	-	-	-	-	76.0
Illinois[a]	-	573.7	4,942.4	2,615.1	809.6	16,583.8	33,650.4	57,652.2	19,062.0	135,889.2
Indiana	197.5	173.0	3,645.2	4,248.8	3,543.4	4,110.5	10,872.8	5,105.9	2,944.0	34,841.1
Iowa	-	-	-	-	-	-	117.1	-	6,405.4	6,522.5
Kansas	-	-	519.9	519.7	1,038.7	2,070.6	4,148.0	8,287.3	4,153.8	20,738.0
Kentucky:										
West	-	-	1,119.6	162.0	336.3	3,793.6	12,759.3	13,643.3	5,081.3	36,895.4
East	13,639.9	8,491.9	2,286.8	1,658.8	1,158.3	2,154.4	24.7	-	-	29,414.8
Maryland	-	-	-	124.6	191.8	208.2	378.6	56.4	220.4	1,180.0
Michigan	-	-	-	-	-	-	-	205.0	-	205.0
Missouri	-	-	-	-	-	-	6,456.7	20,669.2	51,634.1	78,760.0
Montana	51.2	218.2	205.0	397.2	400.0	175.0	40.0	27.0	591.0	2,104.6
New Mexico	5,212.0	5,474.0	-	-	-	-	-	-	-	10,686.0
North Carolina	-	-	-	-	-	110.0	-	-	-	110.0
Ohio	-	611.0	369.0	2,110.2	2,750.4	7,810.5	9,785.3	10,148.2	8,439.4	42,024.0
Oklahoma	250.6	772.2	825.0	368.1	-	-	577.2	19.1	490.6	3,302.8
Oregon	-	14.0	-	-	-	-	-	-	-	14.0
Pennsylvania	44.0	1,154.4	7,624.4	12,424.9	19,689.5	9,995.6	5,287.6	1,150.5	580.6	57,951.5
Tennessee	3.3	160.9	715.9	258.7	178.2	190.5	219.7	43.8	68.5	1,839.5
Texas	-	-	-	-	7,978.0	-	-	-	-	7,978.0
Utah	8,551.4	13,584.0	-	1,524.9	-	-	-	-	3,997.7	27,658.0
Virginia	1,981.5	6,077.5	1,637.1	-	123.9	-	-	-	-	9,820.0
Washington	898.9	672.1	-	-	-	-	-	-	-	1,571.0
West Virginia	20,761.0	26,710.6	21,819.7	13,290.6	8,496.1	2,491.8	3,147.4	5,949.2	-	102,666.4
Wyoming	6,222.2	6,596.6	-	-	-	-	-	-	1.1	12,819.9
Other States[b]	-	616.0	-	-	-	-	-	-	-	616.0
Total	104,168.4	111,502.6	52,260.1	45,179.5	47,307.0	50,111.9	87,505.1	122,957.1	103,688.5	724,680.2
Percent of total	14.4	15.4	7.2	6.2	6.5	6.9	12.1	17.0	14.3	100.0
Subbituminous coal:										
Alaska	71,115.6	-	-	-	-	-	-	-	-	71,115.6
Colorado	13,320.8	4,908.7	-	-	-	-	-	-	-	18,229.5
Montana	94,084.4	36,728.0	0.5	1,303.7	-	-	-	-	-	132,116.6
New Mexico	38,735.0	12,000.0	-	-	-	-	-	-	-	50,735.0
Oregon	87.0	87.0	-	-	-	-	-	-	-	174.0
Utah	-	-	150.0	-	-	-	-	-	-	150.0
Washington	3,693.8	500.0	-	-	-	-	-	-	-	4,193.8
Wyoming	35,579.7	72,315.6	-	-	-	-	-	-	8.6	107,903.9
Other States[c]	-	4,047.0	-	-	-	-	-	-	-	4,047.0
Total	256,616.3	130,586.3	150.5	1,303.7	-	-	-	-	8.6	388,665.4
Percent of total	66.0	33.6	0.1	0.3	-	-	-	-	(d)	100.0
Lignite:										
Alabama	-	-	20.0	-	-	-	-	-	-	20.0
Arkansas	280.0	70.0	-	-	-	-	-	-	-	350.0
Montana	60,214.5	24,141.6	2,660.9	-	-	464.7	-	-	-	87,481.7
North Dakota	284,129.1	34,987.3	31,581.6	-	-	-	-	-	-	350,698.0
South Dakota	-	2,031.0	-	-	-	-	-	-	-	2,031.0
Texas	-	-	6,902.0	-	-	-	-	-	-	6,902.0
Washington	-	116.6	-	-	-	-	-	-	-	116.6
Other States[e]	-	42.0	-	-	-	-	-	-	-	42.0
Total	344,623.6	61,388.5	41,164.5	-	-	464.7	-	-	-	447,641.3
Percent of total	77.0	13.7	9.2	-	-	0.1	-	-	-	100.0
Anthracite:										
Alaska	2,101.0	-	-	-	-	-	-	-	-	2,101.0
Arkansas	-	-	-	145.5	286.3	-	-	-	-	431.8
Colorado	-	90.0	-	-	-	-	-	-	-	90.0
New Mexico	-	6.0	-	-	-	-	-	-	-	6.0
Pennsylvania	12,211.0	-	-	-	-	-	-	-	-	12,211.0
Virginia	335.0	-	-	-	-	-	-	-	-	335.0
Washington	5.0	-	-	-	-	-	-	-	-	5.0
Total	14,652.0	96.0	-	145.5	286.3	-	-	-	-	15,179.8
Percent of total	96.5	0.6	-	0.9	2.0	-	-	-	-	100.0
Grand total	720,060.3	303,573.4	93,575.1	46,628.7	47,593.3	50,576.6	87,505.1	122,957.1	103,697.1	1,576,166.7
Percent of total	45.7	19.3	5.9	3.0	3.0	3.2	5.5	7.8	6.6	100.0

[a] From U.S. Geological Survey Bulletin 1136 supplemented by data from Washington Division of Mines and Geology Bulletin 47 and Iowa Geological Survey Technical Paper 4, with adjustments for production and losses in mining through 1964.
[b] Sulfur levels assigned principally from data published in Illinois Geological Survey Report of Investigations No. 35. New study now in preparation indicates substantially lower tonnages of coals in the sulfur range of 2 percent or less than are shown in this report.
[c] Arizona, California, Idaho, Nebraska, Nevada.
[d] Arizona, California, Idaho.
[e] Less than 0.1 percent.
[f] California, Idaho, Louisiana, Nevada.

Reference: U.S. Bureau of Mines, 1966.

4.3.2 Mining Alternatives and Environmental Impacts

Approximately 80% of the coal mined in the U.S. to date has been obtained
by underground techniques; however, the proportion being surface mined
has risen steadily over the past several years to about 50% currently.
The term "surface mining" used herein is understood to include conventional area, strip, contour, and auger techniques. Neither alternative,
surface or underground, is without environmental hazards as indicated in
Table 6.

Table 6: Environmental Impacts of Mining

	Underground Mining	Strip Mining
THE LAND SURFACE	subsidence coal waste piles	surface disturbance
THE WATER SYSTEM	acid pollution solids pollution	acid pollution sedimentation chemical pollution
AIR POLLUTION	burning waste mine fires	dust (during operation) wind erosion (dry climates)
AESTHETICS	base facilities & waste	entire mined area
ECONOMY AND HUMAN USE	develops tax base accidents pneumoconiosis increased social costs	reduces tax base increased social costs

It is difficult to predict accurately the amount of future coal
production by surface compared to underground mining. The western reserves are plentiful and low in sulfur; this makes them attractive for
power generation with minimum sulfur pollution and for conversion to gas
and liquid fuels at plants requiring large coal reserve tracts. Since
much of this coal is strippable, a growth in production in the western
states would increase the amount of surface mining. On the other hand,
the location of western coals far from the eastern markets, their lower
heat content and their present nonuseability for coke production would
indicate continued substantial production from midwest and eastern mines
by underground mining methods. A marked improvement in our ability to
economically remove SO_2 from stack gases would retain eastern coal in the
electrical generation market. The electrical industry's apparent resis-

tance to addressing the immense equipment development and installation
task does not make it likely that proposed SO_2 emission standards will be
met before 1980 at the earliest. Other means for removing sulfur are
under active study including liquefaction, gasification, and solvent refining. Coal washing has, of course, long been in use and results in a
significant reduction in pyritic sulfur. All these have associated costs
and environmental impacts.

Many of the environmental impacts of mining are only now beginning
to receive theoretical and experimental attention in this country. From
experience in Europe and from experience in analogous fields such as road
construction we may conclude that there are techniques for controlling
many of these environmental impacts in the future. There are some
methodologies, known to be applicable in the eastern part of the country,
for removing, segregating, and replacing soils, for revegetation and for
controlling effects on adjacent lands and waters. But there has been
insufficient experimental application in this country with regard to
either surface or underground mining of techniques to minimize environmental impacts or to reclaim wastes. Certain problems are not near
solution, even in theory; e.g., the restoration and revegetation of semi-
arid western surface mined lands. Research, application and standard-
setting should be accelerated so we may quickly reach the point where
mining is only undertaken on the basis of known, applicable, and enforceable environmental protection procedures.

There is unfortunately very little domestic experience on which we
can rely. Operators are confronted with a wide range of standards, from
very strict to nonexistent, with enforcement equally variable. Reclamation costs, especially for the few good projects, are considered proprietary which surely leads, perhaps not entirely avoidably, to duplication
of experiments. There is considerable European experience with the
problem, notably in Great Britain, the Netherlands, and West Germany, but
the situations are sufficiently different that it is unclear to what extent
parallels can be drawn. Still, many techniques for rehabilitation are available and should be applied. To this end we make the following recommendations designed to control the environmental impacts of surface mining.

We intend it to be understood clearly that in these recommendations
the potential for rehabilitation should be a matter of central concern to
the development plans. One direct consequence would be the prohibition
of surface mining in areas where rehabilitation is deemed impossible for
whatever reasons.

Recommendations

1. The rate of extraction of coal resources should as a matter of policy
be controlled by the rate of rehabilitation and the ability of the project
to meet on- and off-site environmental constraints.

2. It is essential that proposed surface mining operations be required by the Federal Government to develop pre-plans designed to minimize environmental impacts on man and wildlife during mining, to meet on- and off-site air and water pollution regulations, and to define a timetable for rehabilitation.

We concluded that there is great need for uniformity in the establishment of regulations and procedures appropriate to local situations and in the supervision of compliance throughout the mining industry.

Recommendation

We strongly recommend the establishment of Federal guidelines for the development of regulations and procedures relating to the environmental impacts of coal mining. We also recommend that Federal guidelines be established for supervising compliance. Rehabilitation standards must be sufficiently specific to provide meaningful guidelines for preplanning.

We recognize that the implementation of the above requires knowledge of the land's state prior to mining and continual monitoring during operation. Where they are absent, the methods of measuring and monitoring the appropriate parameters should be developed. In addition, surface mining technology development should be directed toward achieving greater depth capability as a means of both making deeper resources available and reducing the extent of surface impacts.

Recommendations

1. It is strongly recommended that environmental baseline studies be conducted for every proposed mine project, whether on public or private lands.

2. Where techniques and methods are known only slightly or are completely lacking, it is recommended that research precede mining activity on an appropriate scale to accomplish satisfactory rehabilitation.

3. A thorough study of the real costs of rehabilitation of surface mined lands should be conducted at the earliest possible time. The study should include an indication of where the costs should be borne.

4.3.3 Disposition of Mine Wastes

It should be realized that the application of environmental constraints in the preplanning process may well determine whether a particular deposit is deep mined, surface mined, or not mined at all. We expect from the evidence available to us that mining activities will continue to expand in all parts of the nation and at a particularly rapid rate in the West.

We concluded that improvements are necessary and possible in the management of underground mine wastes, in the matter of the health and safety of deep miners, and in the control of subsidence.

The control of waste pile fires, water run-off, and unsightly gob piles might be managed by returning the wastes to void. It will certainly involve expense, but here there is considerable European experience, the applicability of which should be investigated promptly. In addition, recent experimental work by the Bureau of Mines has had promising early results.

Recommendation

Federal standards should be established for control of underground mining wastes. These should include as a requirement the return of waste materials to the void where practical and the rehabilitation of surface-stored wastes to the same standards established for surface mined land.

4.3.4 Subsidence

Subsidence poses only a minor threat to forests or farmland, the overwhelming proportion of the land which is or may be undermined. Subsidence is a problem where buildings or other facilities such as pipelines are undermined. It is estimated that subsidence will not become an actual problem in more than 2% of all undermined areas. For future underground mining this problem can be controlled either by limiting underground mining in areas zoned for residential and industrial development, by requiring the "re-packing" of mines with the waste materials where they undermine pipelines or other structures, or both.

4.3.5 Health and Safety

The matter of the health and safety of the deep miners is critical. For fuels used in electrical generation, the accident rate per kilowatt-hour generated is more than ten times higher for coal than for any other fuel (including uranium). The high accident rates have apparently not been reduced substantially by the Coal Mine Health and Safety Act of 1969. For a number of reasons from various perspectives there was general agreement among the team members that the Act is not as effective as was anticipated.

The underground control of coal dust must be improved. Proper ventilation technology is available and has been for some time; the high incidence of pneumoconiosis is inexcusable. The current and projected levels of "black lung" benefits paid to miners surely indicate the wisdom and urgency of directing appropriate funds and regulatory authority to the problem.

Recommendation

The Coal Mine Health and Safety Act of 1969 should be thoroughly and
promptly reviewed with regard to its effects on the health and safety
of the miners. Its provisions should be vigorously enforced with the
aim of achieving substantial reductions in underground mine accident
rates. The Federal government should support increased research and
development in the technologies of economic underground mining and
improved health and safety.

4.3.6 Federal Leasing Policies

Much of the western coal reserves to which we have repeatedly referred
are on Federal or Indian lands, perhaps as much as 70-75% depending on the
outcome of unresolved legal questions regarding Indian ownership of
mineral rights. Fully one-third of our national forests are underlain
with mineral resources, half of them being coal. We find that Federal
leasing programs in the past have been poorly planned, have lacked inter-
agency coordination, and are unevenly administered. The situation has
become nearly chaotic and both the Bureau of Land Management and the
Bureau of Indian Affairs have been recently identified by the Comptroller
General of the United States for their failure to adequately enforce the
restoration of surface mined lands under their control (Comptroller
General of United States, 1972). There is great need for coordination,
reform, and improvement in the coal leasing programs. This is particularly
true in the enforcement of their environmental impact control provisions.

Recommendation

We strongly recommend the establishment of a Federal land use policy which
would be the basis of the mineral leasing program. The Resources Recovery
Act should be revised to include guidelines for improved, coordinated
leasing policies aimed at avoiding adverse environmental impacts as well
as providing controlled access to federally owned fuel resources.

4.3.7 Coal Utilization and Environmental Impacts

The United States currently produces about 600 million tons annually, a
little more than half of which is used for electrical generation as shown
in Table 7.

Table 7: Coal Utilization (1970)[a]

	Tons (millions)
Electrical generation	300
Industrial heating and processing	90
Manufacture of coke	100
Commercial & residential heating	10
Export	70
Total	600

[a]National Coal Association, 1970

Electrical power production has doubled twice in the past 15 years and will probably double again by 1980 (Federal Power Commission, 1970). This is the primary component of the increase in coal production expected over the next several years to an annual level of 990 million tons by 1985. New and more efficient approaches to electrical generation having less environmental impact are needed.

Coal's fraction of the national energy mix is not expected to change significantly from the present 20% through 1985 unless commercial gasification and liquefaction technologies are developed. The prospects for coal gasification, in particular, are very bright. It now appears likely that economically competitive gasification techniques yielding low-sulfur gas will be available within a few years. If development efforts are in fact successful, the environmental impacts could be substantial in the absence of effective controls. For example, a single plant producing 250 million cubic feet per day of synthetic natural gas (SNG) will require 6 to 9 million tons of coal per year, and 700,000 gallons of water per hour. The Federal Power Commission (FPC, 1972) has projected 36 such plants by 1990. (One pilot commercial plant producing low Btu gas is already operating in the midwest, another in the west, and a full-scale commercial plant is expected to be on line by 1976 in the southwest.) Each plant in the FPC projection would over a 30-year life in a "typical" western site require the mining and reclamation of about 15,000 acres. In full operation the 36 plants would produce 3.3×10^{12} cubic feet per year (in 1990, remember), which is about 15% of our current natural gas consumption, and in so doing would use nearly as much coal* as we now use for electrical generation. However, gasification

*1,760 tons per acre-foot, 16 million Btu per ton, 10 foot seam thickness.

has the promise of substantially lower pollutant emission levels than
direct firing, since potential contaminants (e.g., sulfur) are more
easily removed from the gas.

A variety of chemical by-products would also result and might, if
uncontrolled, be expected both in the water and gas effluents. Techniques
are available for controlling them to meet current standards, but the
scale of the potential operation is staggering. Meeting current standards
could still permit very large annual emissions and possible local hazards.

We recognize that Environmental Impact Statements have been or are
being prepared for two of the several proposed gasification facilities
that are furthest along in planning. It is our feeling, however, that
such statements should give more than cursory consideration to alternatives to the proposed action, alternatives being one of the major points
that the statement is directed to address according to the provisions of
NEPA. For this reason we make the following recommendation.

Recommendation

We recommend that the Federal Government conduct a thorough environmental
impact study of the alternative methods of coal gasification, solvent
refining, and liquefaction, including a comparison of the environmental
impacts of various surface and underground techniques, a study of the
impacts of development scales upon the ecosystem, and the definition of
recommendations to Congress for measures to insure that both the technologies developed and the siting and scale of applications are consistent
with protecting the environment.

4.4 Oil and Natural Gas

4.4.1 Reserves and Production

The United States in 1970 obtained more than 75% of its total energy from
oil and natural gas, slightly more than half of that from oil. We are
producing natural gas at the limit of our current ability and it is not
anticipated that we will be able to affect that production rate very much
even with the Alaskan contribution. Hence, the great interest in coal
gasification, liquefaction, and solvent refining, and in the importation of
liquified natural gas.

By the same token, we are producing oil with the spigots wide open;
we have no spare domestic producing capacity (Oil and Gas Journal, January
1973). And more than 25% of our oil is currently imported. This situation
has generated much high level debate and a quickening of interest in
producing oil from coal and, in particular, oil shale.

There is, of course, the potential for increasing production by
finding more oil and gas. Industry estimates indicate that 48% of our

original discoverable oil in place and 63% of the ultimately discoverable natural gas remain to be found.

However, as M. King Hubbert (Hubbert, 1972) has pointed out, finding the rest will not be nearly as easy as finding the current reserves. The barrels of oil found per foot of hole drilled (finding rate) has been dropping steadily as shown in Figure 1. Augmentation of the reserves to the extent required to replace current imports much less future demand

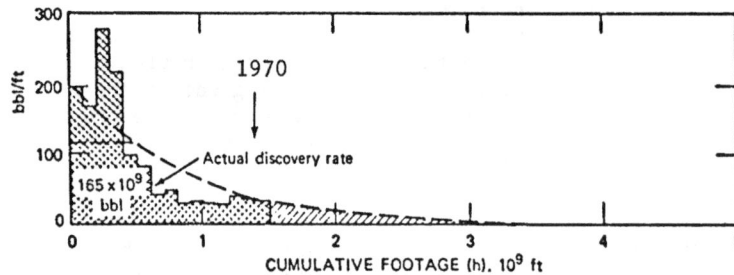

Figure 1: Oil Discovery Rate vs. Footage Drilled

increments by, say, 1980 will necessitate exploration at a remarkable rate. To add another 20 billion barrels of oil to reserves (4-5 years supply at current levels of consumption) will require drilling almost as many feet of hole as has been drilled in this country since the search for oil began.

It is our feeling that, difficult though the task may be, the effort will be made to locate these resources. The usual environmental impacts of oil and gas exploration, random and excessive roads and pipelines, erosions, brine water pollution, and oil spills, will be complicated by the problems of scale of activity. The impacts are controllable, but we must be particularly diligent as exploration expands, particularly in view of the fact that approximately half of our remaining reserves lie in regions of Alaska and offshore. It is absolutely essential that a thorough study of the environmental hazards of oil drilling in the outer continental shelf precede any marked expansion of that activity.

It is also important to point out that much of the known and still-to-be-discovered gas and petroleum reserves lie on Federal lands, particularly offshore. It is our conclusion that the worsening energy picture will lead to their eventual development; however, the existing leasing and enforcement programs have many of the same ills mentioned earlier in the case of coal. The access to these reserves must be carefully controlled with full regard given to potential environmental hazard at every stage of development.

Recommendation

The Federal petroleum leasing program should be promptly modified where
necessary to insure that access, when granted, is in ways that will minimize and confine environmental impacts.

4.4.2 Oil Transportation

The transportation of petroleum presents a special environmental problem
for the United States among the fuel resources because we import such
large quantities. In the process of moving such large amounts of oil,
there is the ever present possibility of spills. Their impact on the
environment is the subject of much continuing research, but there can be
no doubt that it is substantial and there are recent indications that
the impacts may be especially severe when a spill is carried onshore.

While refineries and offshore drilling account for about 25% of the
major spills, at least 44% are due to accidents involving oil tankships,
tank barges, and other vessels. It might reasonably be expected that
increasing tanker traffic associated with increased imports would lead to
an increase in the annual rate of accidents. However, preliminary results
from a recent study of the tanker accident problem indicate that the
accident rate may not increase substantially (Zablotney, 1972). There is,
in fact, the potential for affecting a reduction in the rate and a detailed study of the problem has been undertaken by the United States Coast
Guard under the provisions of the Ports and Waterways Safety Act of 1972.

4.4.3 Oil Shale

The extraction of oil from western oil shales is currently attracting a
great deal of attention as a means of at least partly relieving the
petroleum import situation. By way of maintaining perspective, it should
be noted that the technology has only been demonstrated to be feasible at
the level of 1000 bbl/day, but the size of the industry being envisioned
by the Department of the Interior's Environmental Impact Statement would
provide about one million barrels of high quality refinery feed stock per
day by 1985 or about 5% of our current oil demand and 3% of the demand
projected for 1985 (about equal to the fraction of our current energy
demand that is supplied by hydroelectric facilities). Thus, in terms of
contributing to supply, shale oil will not be a major factor, but it may
be an important component of the 1985 total energy supply.

In contributing to environmental impacts, however, oil shale could
be very significant. An oil shale industry producing one million barrels
per day will require the handling (extraction, crushing, and reclamation)
of about 1.3 million tons of rock per day. By 1985 approximately 20% of
the shale mining might be by surface methods, but whether surface or deep
mined (except in situ retorting which is not considered likely) the spent

shale must be managed in some acceptable way. Its net bulk by virtue of crushing and retorting will be about 25% larger than originally, even after the oil extraction. Consequently, returning the wastes to the void will still leave about one-third million tons per day (at full 1 million bbl/day development) that must be disposed of on the surface. Each mine supplying 100,000 barrels per day (surface) or 50,000 barrels per day (underground) would require as much as 6 million gallons of water each day (see Table 8) and from 60 to 150 acres per year as permanent storage of spent shale (see Table 9).

TABLE 8: Typical Water Consumption for a 50,000 Bbl/Day Oil Shale Plant[c]

Assumptions:

(1) Process sequence includes underground mining, surface retorting, processed shale disposal by wetting and compaction, upgrading of shale crude by partial hydrofining.

(2) Process cooling primarily by aerial condensers.

(3) Raw shale - 30 GPT: processed shale wetted with 20% water for compaction.

	Net Water Consumed Cubic Ft/Sec.
Mining[a]	0.3 - 0.4
Crushing (dust control)[a]	0.2 - 0.3
Retorting[a]	0.8 - 1.0
Processed Shale Disposal[a]	3.4 - 5.8[b]
Shale Oil Upgrading	2.0 - 2.5
Other (Personnel, construction, etc.)	0.1 - 0.6
Total	6.8 - 10.6

---

[a]In situ operation would eliminate most of these water requirements.
[b]If slurry disposal were to be used, this maximum could be as high as 7.0 cu. ft./sec.
[c]Department of the Interior, 1972.

TABLE 9:   Land Requirements for Oil Shale Processing[e]

Function

Mining and Waste Disposal	Land Required, Acres
Surface Mine[a,b] (100,000 bbl/day)	
Mine Development	30 to 85 per year
Permanent Disposal, overburden	1,000 (total)
Temporary Storage; low grade shale	100 to 200 (total)
Permanent Disposal; processed shale	140 to 150 per year
Surface Facilities[c]	200 (total)
Off-site Requirements[d]	180 to 600 (total)
Underground Mine[b] (50,000 bbl/day)	
Mine Development (Surface facilities)	10 (total)
Permanent Disposal	
All processed shale on surface	70 to 75 per year
60 pct. return of processed	
shale underground	28 to 30 per year
Surface Facilities[c]	140 (total)
Off-site requirements	180 to 225 (total)

In situ Processing (50,000 bbl/day)	
Surface Facilities[c]	50 (total)
Active Well area and restoration area	110 to 900
Off-site requirements	180 to 600 (total)

[a]Area required is dependent upon the thicknesses of the over-burden and oil shale at the site. Areas shown are for a Piceance Creek Basin site, with 550 ft. of over-burden and 450 ft. of 30 gallon/ton shale (approx. 900,000 bbl/acre).

[b]Assumes 30 gallons per ton oil shale and a disposal height of 250 ft.

[c]Facilities include shale crushing, storage and restoring (excluded for in situ processing), oil upgrading and storage, and related parking, office, and shop facilities.

[d]Includes access roads, power and transmission facilities, water lines, natural gas and oil pipelines; actual requirements depend on site location. A 60-foot right-of-way for roads requires a surface area of about 8 acres per mile. Utility and pipeline corridors 20 ft. in width require 2.4 acres per mile.

[e]Department of the Interior, 1972.

Disposal of the spent shale is an imposing problem. It might be used to fill semi-wilderness gullies and shallow valleys, with the top soil first having been stockpiled, then eventually placed over the graded spent shale, and finally vegetated. There is the possibility for some shales that alkaline salts may migrate to the shale-soil interface resulting in a root barrier unless covered by deep top soil. The stability of the spent shale disposal piles is improved in some processes through the formation, over a period of several weeks, of pozzolanic type, partially cemented granular particles. Most of the process water used is thereby chemically bound in the spent shale, so the water requirements represent a genuine consumptive use in an area where the amount of uncommitted water is limited.

The effect on the aquifer of the regions is not known in detail. Water table modification is certain, although the extent depends upon the particular case. It is possible that substantial amounts of saline water will result requiring desalting, evaporation or injection disposal.

The study team concluded that oil shale is in some ways unique among the fuel resources. The industry is nonexistent; although total reserves are large, economics project that the ultimate contribution of shale oil in terms of current demand will be very small; unless properly controlled it has the potential for environmental impact out of proportion to its perceptible contribution to the fuels picture; and finally, this is a case where we have a real yes, no, or wait choice regarding development. How manageable the environmental problems might be as the processing plants are developed over the next 30 years is no better known than those for massive stripping of low sulfur western coals. However, it is felt that solutions for these and similar problems typical of the surface and underground mining can be developed and that improved methods for disposing of and vegetating the spoils must be devised. Current vegetation experiments on spent oil shale have yielded less than promising results, but research is in very early stages and is definitely continuing, On the basis of our generally primitive knowl- edge regarding our ability to handle the spent shale problem we make the following recommendations.

Recommendations

1.   It is recommended that the Federal government encourage the de- velopment of the western oil shale reserves by industry only on a prototype basis in accordance with responsible environmental standards.

2.   The rate of extraction of the oil shale should, as a matter of policy, be compatible with the rate of reclamation and the ability of the project to meet on-and off-site environmental standards.

3. It is strongly recommended that any policy governing oil shale development include the requirement that spoil and spent shale be returned to mined-out voids to the maximum extent possible.

After an extended moratorium on shale leasing, the Federal government has now developed a new program beginning with six demonstration leases and has prepared an environmental impact statement for the program. In the demonstration projects there is the opportunity to collect extensive environmental baseline and impact data which should certainly be done as a part of the program. We underscore its importance with the following recommendation:

Recommendation

We urge the establishment of an environmental parameters monitoring system for collection of baseline and impact data as a part of the oil shale lease stipulations.

4.5 Uranium and Thorium

As indicated in Table 1, the nuclear fuels together form a very large energy reserve. However, for the nuclear fuels to assume the 16-17% component of the nation's energy consumption predicted for them by 1985 (National Petroleum Council, 1972; Federal Power Commission, 1970) will require that a number of things happen, not the least of these being the successful demonstration of an acceptable breeder reactor system. Siting problems will need to be solved; the emergency core cooling problem must be resolved; safeguards against plutonium diversion must be devised; excessive reactor down times will need attention; and waste heat must be controlled. It is also worth pointing out that nuclear fuels are electrical generation energy sources and cannot be easily substituted for portable energy sources. We are, frankly, pessimistic that nuclear fuels will be able to assume their projected 16-17% share of the total energy consumption by 1985.

The extraction end of the industry has many of the environmental problems typical of other mining activities plus the obvious additional hazard of radioactivity. The health and accident hazard for miners is certainly real, but the excess mortality rate (all causes) among uranium miners is only about half that of coal miners and in terms of electricity produced is only about 5% (Lave, 1972). Past improper (we know now) disposal of processing mill tailings as construction fill is being corrected and should not be repeated. Tailing piles are now being buried and in some cases the surface must be secured, causing some disruption of wildlife movements. A parameter that should be monitored regularly is the profile of the radioactive contaminants in streams near nuclear fuel extraction and milling facilities.

Recommendation

The rate of extraction and processing of nuclear fuel materials must, as a matter of policy, be compatible with the rate of reclamation and the ability of the project to meet on- and off-site environmental and health standards.

4.6 Hydroelectric

Currently about 16% of our electrical generation requirements (hence, about 4% of our total energy needs) is being provided by hydroelectric sources. Of that 16%, about 1% is pumped storage. The Federal Power Commission (FPC, 1970) has projected that by 1990 the hydro-power percentage will have dropped to about 13% although this represents an increase in the total amount of energy from this source. Of the 1990 projected total, only 7% will be from conventional hydroelectric projects reflecting the fact that nearly all our potential, large hydro sites have been developed.

About half of the 1990 hydro projection and most of the increase in capacity will be due to pumped storage, primarily associated with nuclear generating plants. Such facilities are being proposed for peak load operation at an annual average capacity rate of 20 to 25% as opposed to the 55% rate typical of present pumped systems.

Pumped storage systems for peaking may be environmentally superior to the low efficiency, high fuel consuming peaking equipment now in use (e.g., gas turbines). But, they introduce a new impact, namely the dedication of land to the impoundment. The wisdom of such a dedication must be examined in each case. Given that the increase in electrical generation will occur, pumped storage may be an acceptable peak load source. It should not be planned as a part of base load, since the tendency would be to construct very large reservoirs thereby blocking out extensive sections of land.

4.7 Nonfossil Sources

4.7.1 Geothermal

This source has been in use long enough in Italy and Mexico that it may not be labeled correctly as unconventional. The tapping of steam and hot water deposits, such as the Geysers Field in California, could under favorable conditions provide about ½% of the nation's energy demand projected for 1985 (National Petroleum Council, 1972). It must be noted that making even that contribution will require a 90-fold increase in output compared to the 1970 geothermal energy production. It is on this basis that we are extremely doubtful that geothermal energy could provide the approximately 7% of our total 1985 requirements as projected by the recent "Hickle report" (University of Alaska, 1972). However, significant attention is currently being directed at the determination of the extent of this resource. There are indications that it may be more widely distributed in the nation than has been supposed. If this is indeed the case, the steam and hot water deposits could provide approximately the same fraction of our energy requirements by the late 1980's as will be provided by conventional hydroelectric sources.

If proposed techniques for fracturing hot rock through which water would be circulated actually work, the potential geothermal energy reserve is much larger. However, the pay-off time for the latter is very long indeed, probably well beyond the end of this century.

It should be recognized that producing electric power from geothermal sources has associated environmental impacts. Many wells have substantial emissions of hydrogen sulfide gas (for which removal technology is available) and the waste water may contain large amounts of carbonates, sulfates, and silicates which make surface disposal environmentally unacceptable,if untreated. Extensive surface subsidence can also result. The industrial development of geothermal sources is proceeding, but without some essential environmental safeguards. It should be governed by the same environmental constraints as other sources, including pre-plan submission.

Recommendation

All geothermal energy development should be required by the Federal Government to develop a pre-plan designed to minimize environmental impacts on man and wildlife and to meet on- and off-site air and water pollution regulations.

4.7.2 Fusion

Research is continuing on the development of controlled fusion energy sources. Our feeling, based on conversations with researchers in the field, is that commercially usable results are still very far in the future, at least beyond 1990. While an operational system would likely not be free of environmental impacts, it is too early to address specific recommendations to them.

4.7.3 Solar

Man's production of food and products from plants and animals as well as electricity derived from conventional hydroelectric sources represents most of our utilization of the present solar energy flux. Solar cells have received much past attention. Although they produce expensive electricity, our position may become such that no source can be ignored. Most untested suggestions for utilizing solar energy directly share the common obstacles of relatively low ultimate temperatures requiring new generation technologies and the need for very large land areas. Of particular promise is the use of solar systems for residential and commercial heating and cooling in many areas of the country. This technology should be vigorously pursued.

Recommendation

Parallel with encouraging vigorous research directed at utilizing the solar energy flux, we recommend that the Federal Government establish a coordinated program whose intent would be to anticipate and evaluate the possible environmental hazards.

4.7.4 Additional Sources

Finally, it must be observed that man is an ingenious creature. Besides the energy sources we have mentioned in the course of this report, we are aware of work underway on other fuels and energy sources such as hydrogen fueled systems and solar systems, both ground-based and orbital. The tides and winds are energy sources that have received much attention over the years, sometimes with success, and are the object of renewed interest. None of these nor, it is safe to say, others that we may devise in the future are free of environmental implications. A concern for controlling environment should be an integral part of the development of all future energy sources.

Recommendation

The Federal Government should vigorously support research directed at developing all potential energy sources.

4.8 Recapitulation of Policy Criteria

1. Establishment of a coordinated policy for development of energy fuels by the federal and state agencies in order to meet national goals in protecting environmental quality is of imminent importance.

2. Federal energy resource leasing policy should be based upon sound regional land use planning, since about 50% of the oil and gas, 40% of the coal, 50% of the uranium, and 80% of the geothermal energy resources are on public lands.

3. The Federal Government should take steps to develop ethics in conservation of energy.

4. Pricing policies should be reviewed in light of their effects upon the consumption of energy, and demand studies that relate the quantity of energy consumed to its price should be supported.

5. There is urgent need for a balance in energy development and protection of environmental quality.

6. There is urgent need for vigorous, coordinated Federally supported research and development programs directed at all potential energy sources regardless of whether their individual contributions are currently perceived to be substantial. This group includes the direct solar energy, the winds, the ocean thermal gradients, the tides and others that man may in the future identify.

7. Rehabilitation standards will, to a large degree, determine whether or not the fuel materials will be surface or subsurface mined or not mined.

8. Rehabilitation standards should be sufficiently specific to serve as guidelines for reclaiming disturbed surface and spoil areas. This can serve as a guide for federal agencies dealing with fuel production, for the overall fuel industry, and for the environmentally concerned public.

9. It is recognized that both deep and surface mining will continue to expand to meet future needs.

10. Health and safety of underground mines is of paramount importance and should be a concern of both the mining industry and the national government.

11. Review of present federal and state laws to determine the actual application and enforcement of health and safety regulations to evaluate the effectiveness of protecting the health and safety of miners should receive immediate attention.

12. In all cases where possible, it is believed that solid wastes from underground mining should be placed back underground.

13. The quantity and quality of the fuel materials available can allow for the preservation of some surface areas in a natural state where unusual historical, geological, or scientific value exists and of those areas where rehabilitation is not possible.

14. Goals for improvement of environmental policy should consider full employment goals, national energy needs and consumptive use policies.

References

Comptroller General of the United States, 1972. Administration
of Regulations for Surface Exploration, Mining, and Reclamation
of Public and Indian Coal Lands. (Washington, D. C.: U. S.
Government Printing Office), 36 pp.

Federal Power Commission (FPC), 1970. National Power Survey.
(Washington, D. C.: FPC), Part I

Federal Power Commission (FPC), 1972. Natural Gas Supply and Demand
1971-1990. (Washington, D. C.: Bureau of Natural Gas, Staff
Report No. 2, FPC)

Hubbert, M. K., 1972. "Man's Conquest of Energy" in The Environment
and Ecological Forum 1970-1971. (Washington, D. C : U. S. Atomic
Energy Commission), pp. 1-14

Lave, L. B., 1972. "The Health Effects of Electricity Generation from
Coal, Oil, and Nuclear Fuels" presented at the Sierra Club
Conference on Environmental Effects of Electricity Generation.
(Pittsburgh: Carnegie-Mellon University), 22 pp.

National Coal Association (NCA), 1970. Bituminous Coal Facts.
(Washington, D. C.: NCA), 91 pp.

National Petroleum Council (NPC), 1972. United States Energy Outlook.
(Washington, D. C.: NPC), 179 pp.

Office of Emergency Preparedness (OEP), 1972 The Potential for Energy
Conservation. (Washington, D. C.: U. S. Government Printing
Office), 211 pp.

Oil and Gas Journal, 1973. "Forecast/Review," January, pp. 95-110

U. S. Atomic Energy Commission, 1970. Potential Nuclear Growth
Patterns. (Washington, D. C.: USAEC 1098, U. S. Government
Printing Office)

U. S. Bureau of Mines, 1966. Sulfur Content of United States Coals.
(Washington, D. C.: U. S. Bureau of Mines Circular 8312, U. S.
Government Printing Office)

U. S. Bureau of Mines, 1970. Mineral Facts and Problems. (Washington,
D. C.: U. S. Bureau of Mines Bulletin 650, U. S. Government
Printing Office), 1291 pp.

U. S. Bureau of Mines, 1971. Strippable Reserves of Bituminous Coal
and Lignite in the United States. (Washington, D. C.: U. S.
Bureau of Mines Circular 8531, U. S. Government Printing Office),
148 pp.

U. S. Bureau of Mines, 1972. Mineral Industries Surveys: Coal--
Bituminous and Lignite in 1971. (Washington, D. C.: U. S.
Government Printing Office

U. S. Department of the Interior (USDI), 1972 Environmental Impact
Statement for the Proposed Prototype Oil Shale Leasing Program
(draft). (Washington, D. C. : USDI), vol.1

U. S. Department of the Interior (USDI), 1972. United States Energy
Through the Year 2000. (Washington, D. C.: USDI), 80 pp.

U. S. Geological Survey, 1972. Energy Resources of the United States.
(Washington, D. C. : U. S. Geological Survey Circular 650,
U. S. Government Printing Office), 27 pp.

University of Alaska, 1972. Geothermal Energy, final report of the
Geothermal Resources Research Conference, Battelle Seattle
Research Center, September 18-20, 1972. (Fairbanks: University
of Alaska), 95 pp.

Zablotney, J., 1973. "Marine Collisions and Oil Spills in U. S.
Waters." (Staff study for the use of the Study Team on Fuel
Materials, Study on Environmental Aspects of Materials Policy,
unpublished), 17 pp.

CHAPTER 5

STUDY TEAM

ON

ENVIRONMENTAL PROBLEMS ASSOCIATED WITH FOREST PRODUCTS

JAMES S. BETHEL, College of Forest Resources, University of Washington,
    Seattle, Chairman
LEONARD BERRY, Department of Geography, Clark University, Worcester,
    Massachusetts
THEODORE C. BYERLY, U.S. Department of Agriculture, Washington, D.C.
THOMAS DUNNE, Department of Geography, McGill University, Montreal,
    Canada
PERRY R. HAGENSTEIN, New England Natural Resources Center, Boston,
    Massachusetts
HENRY J. VAUX, School of Forestry and Conservation, University of California,
    California, Berkeley
GEORGE R. WEBSTER, Effluent Guidelines Division, Environmental
    Protection Agency, Washington, D.C.
PAUL J. ZINKE, School of Forestry and Conservation, University of
    California, Berkeley
GENEVIEVE ATWOOD, Staff Officer

This report is to provide an evaluation of the environmental quality
implications of the use and management of renewable resources with
special emphasis on timber and forest products as they contribute
to the materials supply of the United States. The international
environmental implications of the use of forest products, especially
the use of tropical forest resources and the impact of the use of
these forests both on the country of origin and on the United States
is discussed. The team's report does not focus on the potential environmental implications of producing other renewable materials because it
judged that the National Commission on Materials Policy will be less
concerned with these materials than with forest products.

The substantial and increasing importance of using the forest for
purposes other than timber production, such as wilderness protection
and recreation, are recognized as having important supply implications
as well as important sociopsychological values. However the team's
task is to present an evaluation of the environmental implications
of supplying timber to meet materials needs. Therefore, the environmental implications of other uses of forest land, such as the impacts
of recreationists on fragile wilderness ecosystems, have not been
recognized explicitly.

The long run environmental impacts of the well managed production
and use of renewable resources are less than those associated with
nonrenewable resources. Most of the residues from forestry operations
are biodegradable. The energy requirements for the manufacture of most
forest products are far less than for the manufacture of suitable
substitutes produced from nonrenewable resources. The problems of
lessening or eliminating pollution at processing plants are likely to
be less intractable for forest products than for mineral products.

It is evident that a significant portion of the nation's material
needs can be obtained in the future from renewable resources without
degradation of the environment. The important environmental problems
which must be recognized in a materials policy are the subject of our
report.

5.1.1 Consumption of Renewable Resources

Each year for the past half century, forest products have made up
approximately 20 percent of the total nonfuel industrial material
requirements of the United States. The total volume of wood used
during this period has been increasing and is expected to increase
in the United States, as well as in other countries, during the
remainder of this century.

The United States has been a net importer of wood for over 50
years. However, net imports have represented a fairly stable fraction
of the nation's consumption since 1950. We expect that trend to continue
for the rest of the century.

Both economic and environmental considerations lead to the conclusion that a materials policy should include recognition of the great regional differences in our forest resources. Substantial areas of forest land in Alaska and the Rocky Mountains are destined to be removed from timber harvesting for environmental reasons while the concentration of timber production in the South and Pacific Coast region seems more likely for biological and economic reasons.

5.1.2   The Forest Land Base

Of the land base of the contiguous states of the United States approximately 30 percent is devoted to forests, 60 percent to agriculture and the remaining 10 percent to other uses. Of the 634 million acres in forests, 494 million are characterized by the United States Forest Service as commercial forest land. The 140 million acres of noncommercial forest land include land so low in productivity as to be inappropriate for the production of crops of trees for materials supply purposes, as well as forest land designated as parks, wilderness areas, primitive areas and the like that are not available for timber production.

Included in the 500 million acres of commercial forest land are another 100 million acres where the overriding benefit of the forest is its influence on the environment. On lands where the dominant value of the forest is for erosion control, recreation, flood control, or similar forest influences, forest product harvesting should be limited to levels and character consistent with these controlling uses.

Fifteen percent of the forest land now defined as commercial is not likely to permit economic timber growing in the long run and includes only 2 percent of the timber growing potential of the country. To some extent, low timber productivity is consistent with high values for other uses of forest land such as scenery, watershed* protection, and wildlife. But of greater importance ecologically, those sites that are least productive tend to be the most fragile, whereas those that are most productive for timber can withstand more intensive use.

Opportunities for increased timber production in view of environmental management imperatives vary considerably from region to region. For example, timber production from federally owned forests is a major element in the economy of southeastern Alaska. To attract a processing industry, a large part of this productive forest has been committed to timber harvesting in sales contracts running several decades.

---

*As used in this report "watershed" refers to the entire basin that furnishes the flow of water into a stream.

Because so much of the productive forest land of southeastern Alaska is now under these long-term contracts, the opportunity to cope adequately with emerging environmental problems, both aesthetic and ecological, is severely limited. Similar conflicts between environmental and economic needs may not be so clearly drawn in other regions of the United States, but the Alaska experience dramatizes the importance of considering all alternative land uses and their environmental impact before making long-term commitments of forest lands. There should be intensive investigation of the possibilities of minimizing environmental impacts under current contracts.

We expect that the total land area in the United States now in forests will, and should, remain nearly constant in the future. Some of this forest land is not now available for timber harvesting and we feel that additional areas are likely to be withdrawn from timber harvesting in the future. It is likely that the continuing modest shift of submarginal agricultural land into forests will be balanced by shifts of forest land into agricultural and urban uses. We see no urgent reason to convert major areas of land in the United States either from or into forests.

5.1.3 The Timber Growth-Removal Balance

The relations between the annual growth of timber in the United States and annual removal for commodity production and other purposes are shown in the following table. Data for the period 1952 to 1970 are from the Nationwide Forest Survey; the projections of future growth and removals are those most recently prepared by the Forest Service.

The figures show that timber growth has increased significantly over the last two decades. Indeed, the increase of almost 20 percent in annual wood output between 1952 and 1970 has been more than offset by a somewhat larger percentage increase in growth. Recently the level of growth of growing stock (trees 4 inches in diameter or larger) has exceeded the level of removal. However, in the critical saw timber sizes (trees 12 inches in diameter or larger), and particularly in the softwood species which provide the bulk of the wood material for construction and pulping, the record is less favorable. For all species, removals have exceeded growth by a small margin; among soft woods the excess of saw timber removals over growth has been of the order of 20 percent in recent years.

The warning conveyed by this latter relationship is emphasized by the Forest Service's projections of wood growth and removal for the next five decades. These projections assume future forest management practices at about the level of the best management now in effect, and future demand for wood based on assumptions of continued economic growth and continued significant rise in relative prices of wood. These projections show that, although increases in the level of annual growth realized from present levels of management are likely to continue for several decades, these will not keep pace with prospective increases in the consumption of wood products. As a result the relatively small present annual wood deficit between removals and growth, would get steadily larger in the future, other things being equal.

Table 5.1 Net annual growth and removals of timber in the United States.

Year	All species		Softwoods		Hardwoods	
	Growth	Removals	Growth	Removals	Growth	Removals
	Growing stock (Billion cubic feet)					
1952	13.9	11.9	7.8	7.8	6.1	4.1
1962	16.4	11.8	9.3	7.6	7.1	4.2
1970	18.6	14.0	10.7	9.6	7.9	4.4
1980	19.7	16.9	11.5	11.0	8.2	5.9
1990	20.0	18.6	11.8	11.6	8.2	7.0
2000	19.8	20.6	11.8	12.4	8.0	8.2
2020	19.8	20.4	12.0	12.4	7.8	8.0
	Sawtimber (Billion board feet, International 1/4-inch)					
1952	45.1	52.5	29.5	39.2	15.6	13.3
1962	52.3	50.3	34.7	37.7	17.6	12.6
1970	60.0	62.7	40.3	47.7	19.7	15.0
1980	64.7	66.6	43.8	50.4	20.9	16.2
1990	67.6	71.2	46.5	52.5	21.1	18.7
2000	68.8	76.6	47.9	55.4	20.9	21.2
2020	69.2	74.8	48.8	54.0	20.4	20.8

Source: Review draft of "The Outlook for Timber in the United States," by United States Department of Agriculture, Forest Service. January 1973.

5.1.4 Information Needs

The implications of a shrinkage in the area of forest land which is seen as available for production of timber and of a steadily growing gap between long run wood supply and demand suggest that more intensive management of forest lands will be required if current levels of timber production are to be maintained.

Despite the availability of considerable information on the United States timber inventory, there are important gaps in information necessary to guide timber management inputs. There is far more information available today concerning standing timber volume on forest land than there is concerning soil characteristics, land stability and site productivity. Federal policies must give much stronger support to securing better information on site quality and response to management inputs. This information is necessary to insure that timber management efforts are concentrated on the most responsive sites, to minimize the probability that intensive timber management will be practiced on fragile sites, and to provide a basis for improved environmental quality control on timber producing lands.

5.2 Materials Policy and Environmental Policy

Land used for the production of materials from forests also yields other outputs of benefit to man. These include watershed protection, water for industrial and domestic use, recreation services, fish, game, livestock and aesthetic values. Such lands are commonly managed under a multiple use policy. Indeed legislative mandate dictates that most federal and state forest land for which timber production is a management objective be managed also under a multiple use policy. Multiple use management rarely, if ever, involves all possible uses nor does it imply equality among uses. No single use is maximized though some use or uses may be dominant and others may be excluded on particular parcels of land.

The more intensive the use of forest land (whatever the uses may be), the greater the environmental impact. Proper forest management requires optimizing benefit-cost relations, considering all present and anticipated uses. Conflict between forest materials policy and environmental policy emerges when there is disagreement (or ignorance) with respect to costs and benefits whether these are stated in monetary terms or in less tangible terms.

The manner in which land is managed to meet timber supply objectives will influence its capacity for yielding other goods and services. Intensive timber management in a particular forest ecosystem may have great effects on the capacity of the system for production of water, fish, wildlife, recreation and aesthetic values. Often (though by no means universally) these effects are adverse to the levels of nontimber outputs. In order to minimize such negative impacts, there needs to be closer identification than now exists of which lands are to be managed primarily for timber production and which are not.

If, as seems probable, timber production is to be concentrated on the most productive sites, several environmental consequences will accrue. Among these are:

> Timber production forests will in some cases be close to large centers of population and in others will have other important uses, such as recreation. Decisions on the kind of practices that will be permitted or encouraged must be made in the context of overall land use planning wherein all alternative uses and their environmental impacts will be considered.

2.   Increase in timber management intensity on productive
     sites is likely to cause a shift of other uses to less
     productive sites. This could increase the conflict among
     nontimber uses; i.e., hunting, camping and other high density
     recreation uses on the one hand, and wilderness and
     aesthetic uses on the other.

Some environmental impacts directly affect the health and safety
of man or substantially reduce his options for change in forest land
use in the future. The more troublesome environmental impacts, however,
are likely to be those that reduce the long run productivity of the
land base and those that affect human experiences. The most serious
environmental impacts are not necessarily the most obvious. In the
United States a forestry practice that has induced a major long-term
departure from the natural state has been the national effort to
prevent and combat wildfires. Reduction in the area burned annually
and in the intensity of wildfires has substantially modified the
environment. The practice of high grading for the most valuable plant
species has induced important modifications in the forest environment
that are not always apparent to the casual observer. The practice of
clearcut harvesting results in an environmental modification that is
much more conspicuous, but which may not have such long term effects
as the previous two examples.

Most manipulative events which induce major environmental impact
in forest land management occur as relatively infrequent events over
the growing cycle of trees. Despite their infrequency, their impacts
are great and should be undertaken as a part of a long-term management
plan. Harvesting, woods residue disposal, site preparation and regeneration generally occur within a span of five years out of a rotation
of perhaps 30 to 150 years. Most of the major environmental impacts
occur in this period but some have effects of much longer duration.

Many of the early state forest practices laws and some present
legislative proposals at both state and federal levels are aimed at
controlling environmental impacts by prescribing manipulative events.
There are numerous weaknesses in this approach. Regional and local
variations in environmental conditions cannot be adequately recognized
in blanket statutory prescriptions. Particular manipulative events are
treated separately instead of as interactive parts of a productive
system. And rigidity in the prescriptions discourages innovation
in working with complex ecosystems that require flexibility and variety
in management. Blanket prescription of environmental practices can
lead to land management debacles. But after 40 years of detailed
research we know enough to devise programs of public regulation of
cutting practices that will be both effective and flexible.

5.2.1 Environmental Quality Specifications

Environmental quality specifications in the domain of forest management
have generally taken the form of legislative restrictions on forestry
practices or the more recently developed system of requiring environ-

mental impact statements. While these methods have served a useful purpose in limiting the most obvious of environmental impacts and in permitting public assessment of environmental quality problems they are not sufficient to deal with environmental quality at today's level of concern. Neither of these devices provides either adequate control over important environmental impacts or incentives to management to improve environmental control practices.

A viable environmental quality control system applicable to forest land management must include several essential elements:

1. There must be a set of clearly defined specifications which prescribe acceptable limits on environmental quality in terms that are subject to reasonable monitoring.

2. There must be adequate incentives, penalties, or rewards to assure compliance with the specifications.

3. Authority and responsibility for the management of an environmental quality control system must be clearly prescribed as well as authority and responsibility for surveillance over the manager's environmental quality control system.

4. An adequate monitoring technology must be developed and used by responsible managers and external custodians of environmental quality.

This environmental quality control system for renewable resources would require:

1. National guidelines:

As part of a national land use policy, the federal government should foster a uniform basis for establishing environmental quality control systems on federal, state, and private forest lands. Implementation of such systems on state and private lands should be the responsibility of state government. The actual form of regulation would be different on different classes of ownership.

2. Acceptable quality control systems:

The land owner undertaking to manage forest land for materials production should be required, as a condition precedent to his operation, to prepare and file with an appropriate public body an acceptable environmental quality control procedure, identifying the environmental impacts that are to be controlled; specifying quantitatively the acceptable quality level that will be met by the program, indicating the type and extent of monitoring, and providing for the maintenance of appropriate environmental quality control records that would be available to those responsible for regulatory surveillance.

Underlying the development and implementation of a materials policy must be a commitment on the part of forest managers, public and private, to the importance of applying environmental quality control in all phases of timber production. It is an axiom of industrial quality control that "you can't inspect quality into a product, you must build it into the product." The same general axiom applies to forest products supply operations. Environmental quality control efforts must be a fundamental part of the operations themselves.

The most effective quality monitoring technology now available is related to point source pollution. This technology is adaptable to forest product manufacturing operations but is not useful in monitoring most field forestry operations. If rapid progress is to be made in this area of activity it is important to develop new technology for monitoring non-point discharges. It is essential not only to develop inspection and testing methods appropriate to non-point discharges but to establish the locus of monitoring.

For many types of environmental impact the concept of the critical boundary is useful. Critical boundaries include those boundaries between the area involved in a timber operation and such external areas as the land of adjacent owners, rights of way of adjacent public roads, and boundaries of lakes, streams or estuaries. In the case of aesthetic impacts a critical boundary may not be adequate and some relationship to an impact locus would be desirable.

The concept of critical boundaries has long been used on public forest lands to protect stream banks and scenic corridors along highways. In some cases, it is the basis for state regulation of forestry practices. It should be given careful consideration in the development of guidelines for land use regulation by state and other regulatory agencies. The concept of adjacent ownership defining critical boundaries is insufficient in the case of extensive ownerships, especially public forest lands, where a forestry practice may have substantial environmental impacts that are confined to a single land ownership. Further, the concept of critical boundaries loses some relevance in the case of impacts on wildlife. Hard pesticides or chemicals may enter the food chain of animals without crossing a critical boundary.

The policy of leaving regulation of land use to town or county government has generally been followed in the past, although there have been some significant exceptions where the states have adopted forest practices laws. It is evident, however, that regulation of private land use by the states is imminent in much or all of the country. The concept of critical boundaries should be adopted as one useful approach where private forest lands are involved.

5.3 Environmental Impacts Associated with Forest Practices

Since forest management practices are the principal sources of
environmental impacts, any satisfactory environmental quality control
system must be based on the environmental changes that are induced
by these events. The major forest manipulation events related to
renewable resource supply are road construction, harvesting,
intermediate cutting, site preparation, managerial use of fire or control
of fire, tree residue management, forest fertilization, weed control,
pest control, and disaster salvage operations.

Fire is used as a forest management tool in a variety of ways.
Properly prescribed use of fire may be the best available means of
reducing otherwise severe fire hazards in some areas. It is normally
applied as a part of the harvesting, site preparation, and regeneration
regime. Since natural wildfire has an ecological role in maintaining
some desirable pioneer species on forest sites managed fires are
sometimes used to simulate the natural wildfire in site preparation.
The slash remaining after harvesting is frequently burned to reduce
the hazard of wildfire and to facilitate planting of the new crop.
Fire used for silvicultural purposes results in the release of nutrients and sometimes their movement from the site through air or water
flushing systems. The smoke from silvicultural fire introduces
particulates into the atmosphere. Fire used as a forest practice
at harvest time may significantly increase the rate of erosion if
applied where steep slopes and erosive soils are common.

Some forestry practices utilizing even-aged management may result
in a forest of essentially one species. In the agricultural sense
even-aged forestry as practiced in the United States does not result
in a monoculture. Even-aged stands of such species as Douglas fir
and the southern pines are simulations of the natural even-aged stands
which result from forest fires. Even where sophisticated tree improvement methods are used, seed sources are not nearly as genetically
uniform as is the case with agricultural seeds. Wildlings of the
dominant and associated species invariably invade the planted site and
become part of the stand.

The nearest approach to agricultural type monoculture in forestry
might be such large plantings of exotic species as Pinus radiata
in Australia, New Zealand and Chile, teak in Java and eucalyptus in
Vietnam. Even here, the plantations have not exhibited the problems
faced in a clonal monoculture of, for example, the banana type. The
United States has never needed to develop such pure plantings of
exotic species on a commercial basis and there is no evidence
that this is forthcoming.

Damage to the forest environment is an area of legitimate public
concern whether it occurs on public or on private lands. The study
team determined that the long-term productivity of the land was of
higher priority than meeting more immediate demands for timber or
any other forest product. There is evidence as well as public

concern that improper applications of some current timber management practices result in extensive damage to the forest environment either through reducing the productivity of the land or because floodwater, sediment, or pesticide residue ultimately find their way into public waterways even as far as coastal estuaries and the open sea. Forest management practices should not be used where they produce such consequences.

Rapid improvement in environmental quality control is likely to be achieved if restrictions and specifications are related to the environmental impacts themselves. Environmental impacts associated with improper applications of forest practices include:

1. Quantity and timing of water runoff;

2. Soil disturbance, sediment production, landslides, and massive erosion;

3. Changes in stream temperature;

4. Changes in chemical water quality and the relationship to soil fertility; and

5. Herbicide, pesticide and fertilizer effluents.

5.3.1 Influence of Forest Practices Upon the Quantity and Timing of Water Runoff

Timber cutting increases water yields because evapotranspirative losses are reduced when trees are cut down. The evidence suggests that the extent and type of increase is a function of soil characteristics, species of tree cover, slope, aspect, quantity of precipitation, form of precipitation, position of vegetation in relation to water table, and pattern of storm occurence. Silvicultural practices that lead to even-aged management and that include a period of complete stand removal have a greater influence on short-term water yield than do those practices that involve uneven-aged management and only partial cutting. Reforestation usually decreases water yield and converting vegetation from one species to another also causes significant changes of flow.

It is possible that under a careful program of management valuable flow augmentation could result from timber production. But significant flow augmentation will require such measures as temporary elimination of riparian or other vegetation, with consequent serious hazards of soil disturbance and water quality change. In addition, the large increase of flow obtained during the first year after cutting is not sustained. The "extra" yield declines with time so that often significant gains are only realized for a few years. In most commercial logging operations there are few effective incentives to minimize the damage to water resources. Thus, such damage may often outweigh incidential benefits of increased runoffs.

The production of flood runoff is also increased by deforestation.
Field studies of the influence of deforestation on flood hazard are
mostly limited to small storms. In these events on headwater watersheds,
volumes of storm flow increased by less than 10 percent to more than 280
percent of pre-cut values for similar rainstorms (Lull and Reinhart, 1972;
Hewlett and Helvey, 1970). Field studies of the various processes of
storm runoff production indicate that in major storms comparable variations
with perhaps even larger maximum increase are likely to be experienced.

A stream channel is approximately filled by the mean annual flood.
If the mean annual flood is increased following clearcutting, the
increased discharge is accomdated by increased size of the channel.
Most of the accomodation is made by increasing the channel width, implying
extensive bank erosion, and possibly undercutting of sideslopes, roads,
or other structures.

5.3.2 Soil Disturbance, Sediment Production, Landslides, and Massive Erosion

Various forest manipulations may induce substantial movement of soil and
associated nutrients. The extent to which movement occurs is a function
of soil characteristics, climate regime, slope and vegetation. For
example, the stripping of forest cover and the burning of vegetative
residues can accelerate sheet erosion, gullying, and landsliding. Such
soil disturbance with attendant accelerated erosion, and increased transport of sediment in streams occur as a consequence of natural disturbance
such as fire, windthrow, disease and insect attacks as well as by some
timber harvesting. Clearcutting and other timber harvesting methods may
substantially accelerate these processes. The amount of acceleration is
partly a function of topographic, soil, and precipitation conditions and
partly a function of the total extent of initial disturbance, whether by
tree cutting itself or by construction of timber extraction roads.

Increased erosion and sedimentation can result in:

1. The removal of topsoil with its supply of absorbed plant
    nutrients.

2. Deposition of sediment in channels causing a reduction of
    the capacity to convey floodwaters.

3. Changes in channel geometry and pattern.

4. Damage to sport fisheries. Excessive turbidity in stream
    water favors the propagation of less valuable fish over
    game fish because the excessive silt adversely affects the
    gravel spawning grounds of some species of anadromous fish.

5. Aesthetic values. Eroded slopes, clearcut areas and damaged
    stream channels are eliciting complaint from an increasing
    number of people.

6. Downstream damage, whether by accelerated sedimentation in
    resevoirs, ship channels, or biologically productive estuaries.

Haul roads, skid trails and landings are critical as potential
sources of soil damage. Modern management practices result in 3-10
percent of the logged area being occupied by primary haul roads, while
the area disturbed by skidding and loading is two or three times more
extensive than the primary transport network. Whenever sediment production per mile of road per year is high, these areas are obvious sources
of debris. Where roads are a major cause of soil slides, forest practices
should be modified to reduce road mileage. Skyline and high-lead yarding
methods require fewer miles of road than do tractor yarding methods for
the same treatment area. Newly developed balloon and helicopter harvesting
methods still further reduce required road mileage.

In some areas roads cause accelerated mass wasting on steep slopes.
Poor road drainage, addition of road fill, side casting and oversteepened
backslopes are the main cause of mass failure associated with road
building. The processes have been reviewed by Swanston (1971). Several
field studies have shown that landslide frequency is accelerated over
natural rates by a factor of several times to several hundred times in
logged areas with roads (Swanston, 1971). Slides may be induced during
periods of heavy precipitation when roads are built on steep, unstable
slopes and improperly constructed with respect to drainage. Careful
attention to road construction standards with particular preference to
water drainage specifications can reduce the incidence of road-associated
soil slides. On some areas with very steep and unstable slopes roads
should not be constructed.

It has been stated that although such failures occur when roads are
first pushed into an area and logging begins, the processes are short
lived and their effects do not persist for more than a few years after
logging. There is evidence that this is sometimes not the case. Landslide activity usually increases for several years after the removal of
forest vegetation as roots decay and the shear strength of the soil is
reduced. Swanston (1971) reported a lag of 3-5 years between logging
and the onset of slope failure in Alaska, while Rice and Krammes (1970)
describe slow deterioration of soil stability over a post-cutting
period of about 14 years in California.

Much of the debris removed from slopes by mass failure comes to rest
near the valley floor, where it supplies large quantities of sediment to
streams for many years. Anderson (1970, 1971) has shown that a major
flood during and after logging in steep unstable terrain in areas of
heavy rain may also have long lasting effects upon sediment production,

causing higher sediment yields for several years afterwards. Such effects were found to be greatest in those logged areas with roads adjacent to streams and with landings in draws.

In addition, it should be remembered that the sediment transport measured at a basin outlet is only a minimum index of the extent of soil displacement and sediment production. Much of the displaced soil travels only a short distance and is deposited as colluvial deposits and alluvial fans at the bases of slopes, or as channel deposits. Such disturbances of the soil are important; even though the soil is not carried to streams for some time, it lies for many years in a position from which it can be easily moved by unusually large storms.

As noted earlier, if other things are equal, clearcutting per se may lead to increased suspended sediment yields in comparison with partial stand removal. However, for a given volume of timber harvest in newly opened logging areas, clearcutting may require less road construction than does partial cutting. For this reason, decisions as to the optimum timber harvesting practice in any given situation must be made on the basis of informed and careful professional judgment influenced by all relevant environmental and silvicultual considerations. One goal of the cutting practice regulations recommended elsewhere in this report is to insure such careful apraisals.

5.3.3   Impact of Forest Practices on Stream Temperature

Some forest manipulations bring about temporary changes in the temperature of the water in streams, lakes and estuaries. The increases in both temperature and its diurnal rate of change vary somewhat with climatic region, and are greatest in the dense coniferous forests of the North-western United States, where during the summer mean monthly maxima have been measured to increase by 10-15° F and annual daily maxima by almost 30° F (Brown and Krygier, 1970; Lantz, 1971). This environmental impact can result in changes in the biota for which these bodies of water provide a natural habitat. For example, increased water tempera-ture in breeding areas is one of several ways in which pressure has been put on the salmon of the northwest. Salmon, like other sport fish, have limited tolerance to variation in the temperature of their stream habitat. These limits have been transcended on substantial areas in the past. Recently, tighter regulations have begun to work well in some states. But the problem requires further technical and administrative attention.

To prevent unfavorable temperature changes (and excessive stream bank erosion) many forest managers follow the practice of leaving strips of vegetation along the sides of the streams, lakes, marshes, or estuaries, though this practice is not universal. Some states have enacted legisla-tion on shoreline management that requires such strips on water boundaries of public concern.

5.3.4   Impact of Forest Practices on Chemical Water Quality

Chemical water quality requires a consideration of dissolved gases and commonly available nutrient elements produced by rock weathering and atmospheric fixation, and taken up as major plant nutrients (N, P, K, Ca, Mg, S). In addition, water soluble and insoluble colloidal organic compounds need to be considered, including both natural products from vegetation and artificial compounds such as pesticides and their carriers.

Most plant nutrients are stored in the soil in various degrees of availability to plants. The remaining nutrients are in the organic compounds of the forest vegetation and forest floor debris and are constantly being exchanged and recycled with the nutrient pool of soil. Under undisturbed conditions, the export of nutrients from a forest ecosystem often is low with respect to total turnover rates. However, the seasonal injection of nutrients and organic matter by leaffall is a vital factor in much of the stream biology.

Harvesting practices upset this nutrient cycling process in several ways. First, they tax the conservation capacity of the soil by the sudden nutrient release from decomposing logging debris which results in a first-year fluctuation larger than the normal seasonal fluctuations in stream ecology. Second, burning of slash returns nitrogen and other volatiles to the atmosphere, and leaves a residue of nutrients in a highly mobile state in the ash. Leaching and erosion of the structureless ash occurs readily, contributing to the chemical content of runoff. Third, the export of logs and bark results in a nutrient loss from the production site, and generates a disposal problem at the processing site which contributes to water pollution.

It has been suggested (Curry, in press) that clearcutting accelerates export of nutrients, particularly potassium and phosphorus, to a rate which cannot be replaced by rock weathering and atmospheric inputs. This assertion is not agreed upon by all scientists, but the issue of alleged long-term degradation of site productivity and water quality on forest lands is fundamental and must be resolved soon on the basis of relevant technical information. Additional studies of changes in water quality resulting from clearcutting are in progress (Fredriksen, 1971).

The problem of accelerated nutrient loss becomes more severe when the regeneration of the forest is not vigorous. A lack of regeneration vigour may result from (a) disease or microclimatic changes (as happened in northern Sweden), or (b) because sensitive soils are severely damaged by logging and fires (as happened in many environmentally marginal and fragile forest ecosystems of southern Europe, Asia, and North America), or (c) because regrowth is deliberately prevented by means of herbicides for purposes of water yields (as exemplified by the studies of the U.S. Forest Service in New Hampshire, Bormann et al., 1967).

5.3.5   Herbicide, Pesticide and Fertilizer Effluent in Water Courses

Herbicides are used in forest manipulations to control brush and to modify species mixes in favor of commerically desirable species. They are also used to control weed and brush growth along forest road rights of way. Residues from such herbicide applications can enter water courses, lakes and estuaries, although most formulations now used in forests have caused only rare examples of short-term toxicity.

Large scale aerial spray applications of herbicides to forest fall on target and nontarget species alike but are usually applied to be species selective. Drift and careless methods can result in the destruction of nontarget species and across critical boundaries.   Much spray application is at ground level and is highly specific with respect to target trees.

Current evidence suggests that herbicide impact upon water fauna is minimal in the concentrations normally entering bodies of water but although they now appear minimal, problems associated with the herbicide residues should be further studied and monitored.

Forest fertilization is a rapidly developing technology used to increase the growth potential of forest land. In the application of fertilizers, some fraction of the chemical ends up off target. Aerial applications result in deposit across critical boundaries. Such introduction of nutrient additives into water courses can improve water habitat for aquatic organisms where streams are very low in nutrient content, or overenrich aquatic habitats and downstream receiving waters with possible subsequent increase in the rate of eutrophication.

As in the case of herbicides and fertilizers, insecticide and pesticide residues can cross critical boundaries and are not selective with respect to nontarget species. These impacts can result in changes in water fauna or impacts on nontarget populations outside of the property boundaries. There are examples of high mortality rates among fish and insects following spraying of small amounts of pesticides on forests (Thut and Haydu, 1971). Field studies should also confront the longer-term problem of the persistence of the non-water soluble degradation products of these compounds. These degradation products are fat-soluble, can persist for long periods in the soil, or can be ingested and stored in the fatty tissue of microfauna. In time they can be concentrated as they are transported through food chains. Forests contain the remnants of some important game populations and are important gene pools for plant communities. A recent article by Wilkes (1972) shows how an important species of plant can be endangered when its wild relatives are eliminated. The dangers are known in agriculture and the same caution should be applied to forest management.

5.4   Environmental Problems of Timber Processing

The annual production of timber in the United States (5×10^{11} lbs.) exceeds that of steel, cement, plastics or nonferrous metals. While

the consumption of solid timber may be anticipated to increase only in
proportion to population, the per capita consumption of paper and board
products, currently 550 lbs., is still on an increase, leveling off
probably at 800 lbs. by the end of the century. Significant quantities
of paper, amounting to approximately 15 percent of the total production
used in paper production are now recycled. However, relatively clean
homogeneous industrial waste paper and newspaper is usually recycled,
and the recovery of low-grade fiber from municipal solid waste does not
appear to offer much economic promise under current systems and incentives.

5.4.1   Utilizable Solid Waste from Sawmills and Plywood Plants

Recent technological advances have resulted in substantial reduction of
wood waste volume from solid wood processing plants. Increasingly, slabs,
edgings and trimmings are chipped and used as raw material for chemical
pulp manufacture. As an example, the kraft pulp industry in Oregon relies
upon sawmill waste for 80 percent of its raw material needs. Chip exports
to Japan are concurrently increasing. Methods to pulp sawdust also have
been developed. Additional outlets for wood waste are particle board
and compressed firelog production.

Regardless of these developments, utilization of wastes still represents a problem for plants in remote locations, and bark as well as dirty
wood waste are not marketable as raw materials.

5.4.2   Disposal of Nonmarketable Residues by Incineration

In the past, it was a common practice to destroy waste from sawmills by
burning in the so-called Teepee burners, but this practice no longer
survives today. This form of disposal gave rise to air pollution as a
consequence of incomplete combustion and fly ash. Incinerator and
properly modified teepee burners can reduce the formation of smoke and
emission of particulates to a level statisfying the anticipated standards.
Portable incineration units of the same type for the direct disposal of
slash at logging sites are being tested. Successful as the direct incineration systems may become in terms of environmental compatibility, the fuel
value of the residues is not utilized in this type of disposal.

5.4.3   Use of Nonmarketable Residues as Fuel by Direct Combustion

While sawmills have depended on incineration as a method of waste disposal,
pulp mills generally recover some of the heat value of wastes in the form
of steam by performing the combustion in so-called hog fuel burners. It
has been demonstrated that more efficient combustion can be achieved by
improved design of the combustion chamber, and particulates can be removed
from stack gases by a combination of a cyclone separation and wet scrubber
electrostatic precipitators.

5.4.4   Conversion of Waste to Liquid and Gaseous Fuels

Under Section 208 of the "Resource Recovery Act of 1970" a number of
techniques for commercial scale demonstrations of waste disposal methods
have been initiated. Pyrolytic methods, to convert waste to fuel oil
have so far received the most extensive research support. The rationale
for selecting this particular approach is the versatility of storage
transporation and combustion of the resulting non-sulfur fuel oil. However, this process has weaknessess.

The possibility of total gasification of wood waste to fuel gas by
high temperature pyrolysis also seems to have significant potential for
solving wood waste pollution problems.

5.4.5   Atmospheric Emissions from Wood Producing Industries

In addition to problems associated with the combustion of solid wood waste,
discussed in the previous sections, plup mills are facing problems in the
following areas:

1.   Sulfur dioxide emissions from sulfite pulp mills are often
     on an undesirably high level. There is no reason to anticipate,
     however, that emission problems of this nature will not be over-
     come by more efficient scrubber installations.

2.   The emissions of odorous sulfur compounds together with volatile
     phenols from kraft pulp mills form a difficult problem which has
     not been solved in a satisfactory manner. While substantial
     progress has been made in the containment of these malodorous
     emissions, no existing kraft mill can yet claim freedom of periodic
     fouling of surrounding atmosphere. Since the emissions take
     place at several locations of the mills, control measures are
     both complex and costly. As a consequence of this situation,
     efforts to develop intrinsically pollution-free pulping processes
     have emerged recently. Of these, methods to replace sulfur by
     gaseous oxygen in the pulping process have shown promise in the
     laboratory. This research should be exploited to the pilot
     plant stage.

5.4.6   Aqueous Wastes from Wood Processing Industries

The major source of contaminated water from wood producing industries is
the waste water discharged from pulp and paper making operations. The
industry annually produces about 130 million tons of product which requires
an average of over 30,000 gallons of water per ton to process. These
requirements vary widely among different pulping processes.

This contaminated water receives certain types of treatment before
discharge, or will receive such treatment on completion of required
facilities. Generally, primary treatment is used, resulting in a wet low

grade sludge presently disposed of by burning in hog fuel or power boilers. The water then receives secondary treatment, usually in an aerated lagoon, where BOD reduction is accomplished. The result of these treatments is a discharge water stream which will usually meet present water quality regulations. Further treatment in order to control color, foam and other unsatisfactory properties is currently being demonstrated and evaluated. However it should be noted that at the present rate of conversion it will be several years before all mills in the United States attain this higher level of water quality.

The Federal Water Pollution Control Act Amendments of 1972 require that regulations relating to the discharge of pollutants from the pulp and paper industry be promulgated by 1974. It is envisaged that these regulations will require most mills to expand or add on treatment or control techniques to further reduce pollution levels.

The major current need is intensive study to develop ways to reuse water within the process to reduce the present water requirements. Within plant recycle of process water combined with a variety of clean up steps is the best current prospect for reducing water pollution from this industry. There are prospects from present research efforts for significant reductions up to elimination of this requirement. Research to develop internal water recycling methods and control technology seems to be much more promising at this time than the continuation of massive efforts to provide more external water treatment.

Other wood processing industries also have water discharge problems. Plywood manufacturers can now, through careful management, recycle waste water to adhesive make-up and close the loop completely resulting in no discharge. Hydraulic debarkers used in a variety of manufacturing operations present some difficulties. A closed loop recycling system has been developed and has performed satisfactorily thus offering promise for solution of this problem.

Another area of concern, especially on the West Coast, is log storage in water. Pond, stream, or estuary storage results in discharge of residues into the water system. Since this form of log storage normally involves relatively large bodies of water the pollution problem is minor. Discharges are nontoxic. Where this is important dry storage is usually a viable option.

Most of the water quality problems associated with wood processing can be solved using presently available technology. The task of conversion in the pulp and paper industry is enormous just because of the size of the operations. In the case of old mills the cost may not be justified on an economic basis. The major area where new technology is needed is the field of internal water recycling.

5.4.7 Primary Recycling

Primary recycling or the return of the residuals to the forest ecosystem is a method for disposing of residues that has much promise for an industry whose residues are almost entirely biodegradable. The so called dirty residues (e.g., ink, food particles, grease) that are not successfully reprocessed in the secondary recycling of paper are suitable for primary recycling back to the land. There are some unsolved problems that need to be studied before primary recycling is feasible on a substantial operating basis:

1. Different forest ecosystems have different capacities for assimilation of mill residues. The behavior of ecosystems acting as recipients of forest industry residues needs to be studied and ecosystems classified according to ability to receive residues.

2. A major problem in the utilization of primary recycling is that of residue dispersal. Costs of primary recycling need to be compared to costs of secondary recycling or costs of other forms of residue treatment and disposal.

3. Some materials included in pulp and paper mill residues in very small amounts (synthetic resins, talc, clay, etc.) are not biodegradable. Industry experience indicates that, with proper dispersal, these materials can be assimilated into the forest ecosystem satisfactorily. Additional research should be undertaken to study the fate of these materials and their pathways of movement within the ecosystem.

4. The general problem of inadequate technology for monitoring non-point impact emerges in decisions between primary and secondary recycling. Any residues from secondary recycling would have point discharges whereas residues, if any, from primary recycling would generate non-point discharges. Since adequate technology for monitoring non-point impact is absolutely essential to the development of environmental quality control in forestry operations in any case this technology will be available for monitoring primary recycling.

Despite the difficulties associated with primary recycling it is sufficiently promising, given the economic disappointments of applying secondary recycling and the problem of disposing of residues in other ways, to warrant considerably more study. This judgment is reinforced by the fact that there is much interest nationally in using forest ecosystems as residue sinks for municipal sewage and nonforestry industrial wastes. These wastes are much more difficult to deal with since the residues tend to be much less natural to the ecosystem. If the problem associated with land dispersal of forest products residues can be managed, guidelines for land dispersal of other residue may emerge.

5.5 International Forestry

Most forested lands of the temperate zone of the world are committed to use. The major opportunities for expanding the forest resources base through the use of virgin forest lands are in the tropical regions. Where such opportunities exist outside of the tropics they are, for the most part, in low productivity forests that border on the Arctic and Anatarctic zones.

Major areas of original tropical forests are being harvested to convert the land to agricultural use or to use the naturally grown crop of timber. As in temperate regions, management of native species does not generally occur as long as there are large areas of natural forest in the area. When there is a desire to change the species mix to exotic species such as teak in Java or eucalyptus in South Vietnam or Monterey Pine in Australia, New Zealand and Chile, managed forests result from the establishment of plantations. In some former colonial areas where particular species were in great demand and supply was obviously limited, some forest management for native species has been introduced. An example of this was the development of a managed mangrove culture on the Camau peninsula of South Vietnam.

The intensive exploitation of natural forests from tropical areas has accelerated in many areas of the world that previously were lightly used. These include some areas of Asia such as the Phillipine Islands, Malaysia and Indonesia and some areas of Latin America such as Brazil, Colombia and Ecuador. The principal impetus for this exploitation has come from highly industrialized forest land-poor countries such as Japan, Israel and some European nations. The United States has imported some products of these exploitations but has been a relatively minor user. These woods are not currently needed to meet critical wood supply problems in the United States, but are used because they are less expensive in some instances and have desirable properties not possessed by domestic woods in other instances. U.S. firms are actively engaged in the exploitation of these woods as managers or partners in the management of the conversion activities, but are in the minority among conversion firms.

Even though the United States is a net importer of woods it is also an exporter of wood. Many nations having large forest inventories need long fibered softwoods that they cannot grow. Some wood residues and low quality logs not now saleable on the American market can be sold on the world market. It is sometimes proposed that as a forest conservation measure export of wood ought to be prohibited. It can also be effectively argued that free exchange of wood in the world market to allow for greater efficiency in wood utilization will contribute to better world-wide forest conservation and a more favorable situation with respect to the environmental consequences of using forests for timber production.

5.5.1 Environmental Impacts

There has been considerable concern expressed in many places, including the Stockholm conference, about forest exploitation and its environmental consequences in the tropics. The conversion of natural forests to managed forests or to other land uses always constitutes a major environmental change. Uncommitted land is probably going to be committed to long-term use in the underdeveloped tropical countries of the world much as it has been in the developed temperate countries with decision in the hands of the governments and people soverign in the locale. The opportunities for extranational or international involvement in this decision making process are probably limited, but one opportunity exists where extra-national or international organizations are involved in the land use development either through provision of capital or technical assistance on an intergovernmental basis.

It is appropriate to consider the environmental consequences of using currently uncommitted tropical forest lands to meet world wood supply problems particularly if future wood supply problems internal to the United States encourage this nation to enter the world market in search of these woods in a substantial way. We have identified a number of environmental concerns with respect to the exploitation of natural forest of the tropics.

5.5.2   Loss of Productivity

The harvesting of natural forests may result in loss of the productive potential of forest lands and thus foreclose future opportunities to use the land for managed forest crops or for other land uses. Forested land has generally survived the harvest of the original forest stand in such a status as to permit its future use, but the major world experience with this problem has been in temperate regions and under different technologies. This experience may not be properly transferrable to tropical zones, although the widespread practice of shifting agricultural crops suggests that even very severe disruptions of forests do not prevent their reestablishment at a later date. A forest fallow period is commonly used as a natural restorative of land productivity following such cropping.

Where loss of land productivity has occured in agriculture in the tropics, it has most commonly been the consequence of excessive or unwise agricultural use before abandonment or too short a forest fallow period between argicultural uses. In the case of some lateritic soils, harvesting the orginal forest stand does result in failure in the development of future land uses. The areas where this is a problem are a relatively small fraction of the natural tropical forest--at a maximum perhaps 10 to 20 percent of the total land area. Nonetheless, information concerning soil characteristics is very sparse for the tropical forest areas now being harvested. To the extent that the nations now harvesting their natural forest lands intensively for the first time request help, it is appropriate for developed nations to assist in this sort of soil and land productivity assessment.

5.5.3   Alteration of Natural Habitats

The conversion of natural forests either to managed forests or to other uses alters the environment for animals and plants including those being

fostered in the managed enterprise. The establishment of reserve areas
to provide a residual environment for such organisms is one method that
has been used effectively in many regions of the world.

Little is now known concerning the appropriate number, size and
geographical location of such reserves if they are to be effective.
Countries with relevant information might quite properly provide advice,
guidance, and financial assistance for studies to establish necessary
guidelines when this is requested. Discussion at the V General Assembly
of the International Biological Program suggest that there is considerable
resistance in some tropical countries to the adoption of foreign standards
of land use and environmental quality.

5.5.4   Technical Information Needs

Concern has been expressed that extensive use of natural forests in the
tropics in the absence of adequate knowledge of the structure, properties
and potential uses of the woods might result in loss of potentially
valuable resources. The diversity of the tropical forest coupled with
the lack of scientific research on tropical trees and woods suggest that
this is a reasonable concern. There are few significant research organizations that are engaged in the study of forest tree species in the tropics.
There are far too few significant arboreta, herbaria and wood collections
in tropical countries of the world. The arboretum of Bogor, Indonesia,
is an important exception to this rule. Activities of the sort represented
at this research station should be duplicated many times and extended in
scope. A very few sophisticated ecosystem studies in, perhaps, Asia,
Oceania, Africa and Latin America ought to be initiated--no more than
four such studies in all. Clearly the tropical environments in these four
major regions are so varied that a single ecosystem study in each one
could not be expected to be representative of the region.   Nonetheless
a significant start should be made. Major attention needs to be devoted
to the extension of basic taxonomic studies and the introduction of growth,
yield and utilization research in many areas. These are fields where the
United States and other temperate countries could extend assistance to
tropical countries.

5.5.5   Canadian Forestry

In the domain of international forestry Canada is important as a major
source of United States timber supply. The situation here is, however,
very different from that relating to tropical countries. Canada has a
forest industry that, while smaller, is essentially the equivalent in
structure and quality to the United States industry. Its forest land
ownership is predominatly provincial. Its forest products are inter-
changeable with United States products in the world and United States
market. The same environmental concerns have emerged in Canada with
respect to forestry practices that are discussed generally in this

report. It can reasonably be expected that Canada will manage its
environmental problems as well as the United States and that this may
well have impacts on timber supply from Canada in the long run.

5.6   Barriers to Timber Production

There are barriers to meeting material needs that must be considered
along with the environmental impacts of timber production, harvesting, and
processing. These have been long recognized by foresters, but deserve
special note here because of their significance to a national materials
policy. Moreover certain long run timber supply problems may lend themselves
to simultaneous solution along with more immediate environmental ones.

The major consideration in the production of timber raw material is
the long time period required to grow trees to a merchantable size. A
timber crop generally requires a minimum of two decades to reach merchantable size and may not be ready for harvesting for a century or more
Furthermore, increases in timber volumes require that there be a significant
inventory of standing timber to which new growth is added. The quantity
of standing timber necessary to sustain a given annual harvest ranges from
20 to 40 times the size of the annual harvest.

An important result of this relationship is that there is no short
run timber supply problem in the ordinary sense of a physical shortage.
In general the United States is not likely to be seriously constrained
in its use of timber between now and the end of this century by the size
of the timber inventory. (The present exaggerated level of timber prices
reflects inelasticity of market supplies more than it does sharply
increasing scarcity in the short run.) But over a longer period of time,
the ability of the forest to sustain the present level of harvest, or an
increased level, is dependent on maintaining a very substantial timber
inventory as forest growing stock, as well as on the level of management
applied to the forest resource.

Because of the barriers to reliance on the marketplace in allocating
resources to timber production and in devising programs to ameliorate
environmental degradation, we believe it is important that a materials
policy be devised that will at minimum ensure that there are no unncessary
barriers to environmentally sound timber production. Of major importance
here is the tax structure, the effects of which can negate positive efforts
to meet material and environmental needs.

5.6.1   Lack of Management Inputs

Every evaluation that has been published since 1940 of the long term supply/
demand outlook for timber products has shown that the trend in wood demand
is expanding significantly more radidly than the trend in supply. Consequently, prices for raw wood fibre will almost surely continue to rise
into the indefinite future.

Despite this clear economic outlook, there have been only modest
increases in the amount of effort devoted to managing forest land better.
Much potentially productive forest land is only partially stocked with
the cover needed to produce future timber supply; and much forest land
is handled with little thought for environmental impacts. Even conservative
estimates conclude that we are currently using only 35 to 50 percent of
our available timber growing capacity. It seems clear then that during
the past 30 years existing institutions governing the rate of investment
in forest management have <u>not</u>, in fact, produced a rate of investment in
this renewable resource consistent with either the outlook for national
needs for future timber supply or with current needs for environmental
quality.

In the public sector the principal institutional obstacle to such
investment is the federal agency budget for forest management activities.
Over the past decade budgets for current timber sale activities of federal
forest management agencies have been funded to more than 90 percent of the
level estimated as necessary for balanced multiple use of these lands.
But the level of funding for long-term timber growing activities has been
only 40 percent of the level needed for balance; that for recreation only
47 percent and that for soil and water management only 55 percent. Thus
public land management has been badly balanced. An adequate national
materials policy should correct such biases, which have continually favored
immediate and short term supplies of forest goods and services at the
expense of adequate provision for longer term raw material and environmental needs.

5.6.2   Lack of Incentives for Good Management

The situation with respect to incentives for better management on private
land is far more complex. Timber growing investments on private land
owned by the timber industries (13 percent of the commercial forest area)
have been more intensive than on any other ownership sector. But even
here the record is spotty and there is still substantial idle timber
growing capacity on industrial ownerships.

On private nonindustrial forest holdings (59 percent of the commercial
forest area), the record of investment in timber growing is more dismal
than on either public or industrial ownerships. Moreover, this record
has been compiled during a period when significant public policies were
being applied in an effort to correct the problem.

The average private nonindustrial owner holds only 60 acres of
commerical forest land--a tract too small for independent economic
operation and producing at best insufficient annual income to constitute
a principal source of the owner's support. Such owners may often hold
the property for reasons other than commercial timber production. Those
owners who do consider investing in timber face several discouraging
obstacles. These include a general property tax system which has the
effect of penalizing deferred income investment, an inheritance tax system

which is adverse to long term investment in timber growing, virtual
unavailability of credit for terms commensurate with the length of the
timber investment-harvest cycle, inability to insure the timber growing
asset against fire or other causes of loss, imperfect markets for the
matured timber product in which the small forest owner is at a serious
disadvantage, and rates of return on investment from even the most
efficient forest management which appear to be modest in comparison
with other alternatives offered in the capital market.

If the institutional picture is less than encouraging for private
forest management directed toward timber production, it is wholly negative in encouraging such owners to manage for environmental benefits.
Virtually all of the environmental benefits and costs associated with
forest management are outside the market system. Watershed protection
and erosion prevention values, stream flow modification, wildlife and
fishery values, and most sorts of forest recreation value are not mar-
keted. Thus, the private forest landowner has virtually no incentive
to manage his land so as to enhance these environmental benefits or to
minimize the environmental costs of his actions in these areas. The
situation is changing somewhat as broadly based water and air quality
controls come to bear on forest owners. But that approach is at best
piecemeal and falls far short of achieving the balanced environmental
enhancement that is possible under well-conceived programs of multiple
purpose forest management.

5.6.3 Existing Programs for Nonindustrial Private Lands are Weak

Since 1926 various cooperative (federal/state) programs of education,
technical assistance, subsidies, and tax concessions have been adopted
to try to encourage better forest management on nonindustrial forest
ownerships, particularly with a view to increasing timber growing.
The impact of these programs has been limited at best, and lends little
support to the view that the problems of inadequate management on private forest land can be solved by more vigorous pursuit of existing
policy lines. The combination of a prospectively serious long term
timber materials deficit with the need for much more effective incentives
for achieving environmental benefits through forestry suggests that new
and more effective initiatives in public policy are essential at this
time.

Since nonindustrial private owners hold almost 60 percent of the
productive forest land and are in a situation where economic incentives
to adequate forest management are utterly inadequate, new policy thrusts
ought to be directed at this most critical sector of the forestry economy.
The problem has effectively resisted solution for 50 years so it is
hardly possible to prescribe a definite solution here. But we urge,
that, along with other new policy approaches, serious attention should
be given to establishing public corporations to function as Regional
Forest Management Agencies.

These corporations, with appropriate access to sources of investment capital, would be authorized to lease nonindustrially owned private forest land for at least the length of a timber production cycle for the purpose of managing it intensively both for increase in future timber supply and for current environmental benefits. Industries have successfully used this to enhance their long-term timber supply in the South. It is now timely to see if the concept cannot be adopted to serve the public interests in better forest management to produce raw material supply for tomorrow and environmental benefits for today.

5.7   Recommendations

1.   To the extent that the impacts of timber production prove to be less environmentally costly that those resulting from the production on non-renewable resources for comparable uses, the United States and world capacity to produce these renewable resources should be maintained and enhanced.

2.   An area equivalent to the present total area of forest land in the United States, about 753 million acres (634 million acres of the contiguous states and 119 million acres of Alaska) should be maintained in forest cover in the future, to be managed for all products of the forest including timber, wildlife, water, recreation, aesthetic pleasure and wilderness (including the preservation of genetic diversity).

3.   Because forests require long time periods to grow to merchantable maturity and because environmental quality control should be a part of the planning and implementation of the production process, we recommend that a Federal public land use policy be established.

    a.   Such a policy would require clear identification of those forest lands that should not be included in the long run timber growing base either because the sites are too fragile or because the lands are insufficiently productive to permit economical timber growing. Where such excluded lands are in Federal ownership, timber should be harvested from them only to the extent that is fully consistent with the nontimber management objectives for such lands.

    b.   The Nationwide Forest Survey should be extended and augmented to intensify the inventory of forest productivity, to include other important environmental parameters, and to provide improved information on management response. This vital information is needed for both public and private forest lands.

    c.   Timber harvesting should not take place if the long-term productivity of the site will be degraded or if the long-term costs of regeneration of the timber stand and protection of the site exceed the short-term economic values to be obtained through timber harvest. Forest lands in these categories should immediately be classified as "noncommercial" and dropped from calculations of timber production potentials.

4.   The use of various forest practices, for example clearcutting and chemical herbicides and pesticides, must be carefully planned and monitored. Blanket prescriptions to use, or not to use, such management practices should be avoided. As a component of a national materials policy the federal government should foster a program of environmental quality control that encourages forest managers to incorporate environmental quality into the regular ongoing forest production process. This program should recognize regional environmental differences. The program should be comprised of the following components:

a.   There should be developed clearly defined and verifiable quality specifiations based on impacts on definable environmental characteristics, subject to reasonable monitoring, including impacts across critical boundaries where relevant.

b.   Enforcement should be based upon surveillance over an approved operators quality control system.

c.   Positive incentives for compliance should be developed and instituted.

d.   A major research effort should be undertaken with federal support to develop methods for the monitoring of non-point pollution impact.

e.   Where sites are so fragile that the best known technology will not afford reasonable assurance of protection against deterioation of site or environment, timber harvesting should be deferred until the knowledge requisite to protect the basic resource can be applied.

5.   Because nonindustrial private owners hold almost 60 percent of the productive forest land and are in a situation where economic incentives to adequate forest management are utterly inadequate, we recommend the establishment of public corporations to function as Regional Forest Management Agencies. These corporations, with appropriate access to source of investment capital, would be authorized to lease nonindustrially owned private forest land for at least the length of a timber production cycle for the purpose of managing it intensively both for increase in future timber supply and for current environmental benefits.

6.   Present and future research must be more strategically aimed. We now have a good quantitative idea, significantly fostered by the IBP, of the way in which many variables interact in the major forest ecosystems of the United States. Support for organized forest biome studies should be continued, but increased emphasis in other forestry research must be placed on developing applications of basic research findings to management problems.

7.   The forestry research program should include research devoted to the more environmentally efficient use of forest products, including

alteration if it proves necessary, in the nature and mix of consumption patterns and demand for forest products. It should support research and pilot studies leading to the development of feasible primary recycling technology for biodegradable wood residues. It should foster research devoted to the development of means of converting solid waste from forest products and their derivatives to usable energy.

8.   The United States should, in looking to international trade as a source of forest products and other renewable resources, encourage the production and harvesting of these resources in a manner that is consistent with maintenance of the long term productivity of the land base. We recommend that:

    a.   The United States support programs that will encourage participating countries to value their renewable resources and use them to their best advantage consistent with the maintenance of a permanent resource base.

    b.   As a part of national materials policy the United States should extend assistance to the developing nations of the tropics that are just beginning to exploit their forest resources. This is subject, of course, to a desire on the part of each country for such assistance.

    c.   The United States and international funding organizations that provide financing for forest utilization programs in tropical countries be encouraged to continue to expand their efforts to obtain environmental impact studies as a part of the documentation related to these funding efforts.

References

Anderson, H. W., 1970. "Principal Components Analysis of Watershed Variables Affecting Suspended Sediment Discharge After a Major Flood," Proceedings of the Symposium on the Results of Research on Representative and Experimental Bases. (Wellington, New Zealand: International Association of Scientific Hydrology, Publication 96), pp. 404-416

Anderson, H. W., 1971. "Relative Contributions of Sediment from Source Areas and Transport Processes," Proceedings of a Symposium on Forest Land Uses and Stream Environment. (Corvallis, Oregon: Oregon State University Press), pp. 55-63

Barney, D. R., 1972. The Last Stand, preliminary draft of the Nader Study Group Report on the U. S. Forest Service. (Washington, D. C.: The Center for Study of Responsive Law), 379 pp.

Bormann, F. H., G. E. Likens, D. W. Fisher, and R. S. Pierce, 1967. "Nutrient Loss Accelerated by Clear Cutting of a Forest Ecosystem," Proceedings of a Symposium on Primary Productivity and Mineral Cycling in Natural Ecosystems, Ecological Society of America and American Association for the Advancement of Science. (Orono, Maine: University of Maine Press), pp. 187-196

Brown, G. W. and J. T. Krygier, 1970. "Effects of Clearcutting on Stream Temperature," Water Resources Research, v. 6, n. 4, pp. 1133-1139.

Curry, R. R., in press. "Geologic and Hydrologic Effects of Even-aged Management on the Productivity of Forest Soils, Particularly in the Douglas Fir Regions," Paper presented at the August 1972 Symposium on Forest Biology, Oregon State University, Corvallis, Oregon. (Corvallis, Oregon: Oregon State University Press)

Fredriksen, R. L., 1971. "Comparative Chemical Water Quality of Natural and Disturbed Streams Following Logging and Slash Burning," Proceedings of a Symposium on Forest Land Uses and Stream Environment. (Corvallis, Oregon: Oregon State University Press), pp. 125-137.

Hewlett, J. D. and J. D. Helvey, 1970. "Effects of Forest Clear Felling on the Storm Hydrograph," Water Resources Research, v. 6, n. 3, pp. 768-782

Kitteridge, J., 1948. Forest Influences. (New York: McGraw-Hill), 394 pp.

Landsberg, H. A., L. L. Fischman, and J. L. Fisher, 1963.
    Resources in America's Future.(Baltimore: Johns Hopkins
    Press), 1017 pp.

Lantz, R. L., 1971. "Influence of Water Temperature on Fish Survival,
    Growth and Behavior," Proceedings of a Symposium on Forest Land
    Uses and Stream Environment. (Corvallis, Oregon: Oregon State
    University Press), pp. 182-193

Lull, H. W., and K. G. Reinhart, 1972. Forests and Floods in the
    Eastern United States, United States Forest Service Paper
    NE-226 (Upper Darby, Pa.: U. S. Department of Agriculture), 94 pp.

Rice, R. M., and J. S. Krammes, 1970. "Mass Wasting Processes in
    Watershed Management," Proceedings of the Symposium on Interdis-
    ciplinary Aspects of Watershed Management. (New York: American
    Society of Civil Engineers), pp. 231-259

Swanston, D. N., 1971. "Principal Mass Movement Processes Influenced
    by Logging, Road Building, and Fire," Proceedings of a Symposium
    on Forest Land Uses and Stream Environment. (Corvallis, Oregon:
    Oregon State University Press, pp. 29-39.

The Institute of Ecology (TIE), 1972. Man in the Living Environment.
    (Madison: University of Wisconsin Press), 288 pp.

Thut, R. N., and E. P. Haydu, 1971. "Effects of Forest Chemicals
    on Aquatic Life," Proceedings of a Symposium on Forest Land
    Uses and Stream Environment. (Corvallis, Oregon: Oregon State
    University Press), pp. 159-171

United States Forest Service, 1965. Timber Trends in the United States,
    Forest Resource Report No. 17. (Washington, D. C.: U. S.
    Department of Agriculture), 235 pp.

United States Forest Service, 1972. Forest Statistics for the United
    States, by State and Region, 1970. (Washington, D. C.: United
    States Department of Agriculture), 96 pp.

United States Forest Service, 1973. Review draft of "The Outlook
    for Timber in the United States" (Washington, D. C.: U. S.
    Department of Agriculture)

Wilkes, H. G., 1972. "Maize and its Wild Relatives," Science,
    v. 177, n. 4054, pp. 1071-77

CHAPTER 6

STUDY TEAM

ON

ENVIRONMENTAL QUALITY, BASIC MATERIALS POLICIES,
AND THE INTERNATIONAL ECONOMY

*RALPH C. d'ARGE, Department of Economics, University of California,
    Riverside, Chairman
GARY C. HUFBAUER, Economics Department, University of New Mexico,
    Albuquerque
INGO WALTER, Graduate School of Business Administration, New York
    University, New York, New York
RUBEN S. BROWN, Staff Officer

*Appointed by Dr. Wollman to chair the working group.

6.1 Introduction

Three elements dominate discussions of the international economics of environmental quality and security: (1) The assimilative capacity of the human environment with respect to any given pollutant differs considerably from one region and nation to another. (2) Different cultural groups have their own priorities as to which kinds of environmental abuse are most worthy of correction, and they have their own ideas as to how much effort should be allocated to domestic environmental clean-up or control programs. (3) Societies are not at all consistent about the control techniques which they choose to apply. Some political jurisdictions set standards and issue permits; others explicitly or implicitly vest property rights and establish liability rules; some subsidize the installation of pollution control equipment and the development of pollution control technology; a few even use partial effluent taxes to encourage reduced pollutive discharges.

Of prime importance is the case where environmental abuse is directly or indirectly attributable to a production process--whether basic materials extraction or fabrication--and its consequences are essentially confined to the nation which hosts the productive activity. This is a common case, including such examples as the factory discharge of noxious chemicals into streams or effluents into the air. The preferred economic prescription is generally to make the polluter pay for environmental damage caused. But when the polluter is made to pay either by imposition of a tax or through the purchase of pollution rights, he is naturally concerned that efficient production will shift elsewhere, possibly to competition in other countries. The producers' concern may be well placed, but proposed remedies such as countervailing import barriers are not. If production cannot pay its own way, including losses due to pollution, it is not worth having. Nor should trade restrictions be applied as a result of a nation's false sense of concern about the additional pollution abroad. The foreign country is quite capable of deciding for itself what the best tradeoff is between environmental quality and material income. Compensatory trade barriers represent merely another form of protectionism.

Short-term balance of payments problems that result from a shift of production abroad can be dealt with partially through an adjustment in exchange rates. Problems of structural adjustment can be attacked by means of assistance measures designed to ease the process of change for the affected industries, and by consistency in the standards themselves.

A country may attempt to avoid the trade consequences by subsidizing exports and taxing imports. If a country chooses to deal with pollution by subsidizing the installation of control equipment, then it should live with that decision and foreign countries should accept it. They should not resist the consequent shift of

production or trade, even though the long run global efficiency of resources may be affected. However, international competitive efficiency
will be <u>maximized</u> to the extent that all nations uniformly adopt either
a "polluter pays" principle or a subsidies approach toward environmental
control. Such uniformity is a necessity for efficiency in the use of
global environmental resources.

A second problem area is the case where consumption of some commodity leads to environmental abuse with the consequences largely
confined to the consuming nation. Use of the private automobile is
an important example; also, nonreturnable containers and various nondegradable plastics may be cited. Again, the preferred solution is a
tax, although in some extreme cases the product might be banned outright. More likely is the establishment of design standards which reduce the external harm of certain consumption goods either in normal
use or as a residual. In either case, imports of the offending commodity should be treated in the same manner as domestically produced
goods. Exports should be exempt from any control measures, since the
task of controlling trade in the affected products should be assumed
by the importing countries. Application of this rule concerns all
trade, including raw materials, and will bring uniformity to the control of domestic consumption while at the same time allowing foreign
countries to devise their own unique approaches to consumption externalities.

There is a danger that when standards are used to implement
consumption controls, the standards will be tailored to the design
of domestically produced goods, thereby freezing out imports in the
interest of protection, rather than in the interest of the environment. While this impact is undesirable, the possibility of dis-
guised protection does not require, as a safeguard, prior international approval of national consumption standards. If a country's
producers are sufficiently powerful that they can cynically warp
environmental standards to protectionist purposes, then it is
quite likely that they can devise other nontariff barriers to
serve the same end. The imposition of controls on consumers'
pollution may also give rise to balance of payments and employment problems. Appropriate remedies here include unilateral
exchange rate adjustments and/or national fiscal and monetary
policies.

Finally, there is the problem of transnational environmental
pollution, arising either from production or consumption activities.
There appear to be two categories or types of transnational environmental externalities. In the first category is pollution of the
ocean and estuaries stemming from improper fishing practices, the
discharge of wastes, and in some instances, the emission of airborne
effluents which may, in the long term, affect all countries. In
the second category are one-way pollution flows which originate in
one nation but principally offend certain foreign countries. The

present report considers the second issue only tangentially, since
it has been considered in greater detail by the international environmental policies team (chapter 7). This does not, of course,
reduce the potential economic significance of transnational pollution.

Recommendations

1.   In cases involving pollution entirely within national boundaries,
other nations should not attempt to impose sanctions, tariffs, or
other means to achieve environmental quality beyond that desired by
the individual nation. It is important not to attempt to impose
uniform ambient environmental quality standards among nations. However,
international coordination on emission or product standards may be
useful to preserve harmony among trading nations.

2.   If consumption activity yields domestic environmental degradation,
then it is a matter of indifference whether the good being consumed
is of domestic or foreign origin. Consequently, foreign and domestic
goods should be treated equally in terms of environmental controls,
and deviations from this general principle should be viewed as nontariff barriers to international trade.

3.   All products entering into international trade and potentially
or actually harmful to human health or to the environment should be
clearly marked as such in language understandable at the point of
final use. International standards to this end should be developed
with all deliberate speed.

4.   There is no and should be no justification whatever for trade-
policy measures designed to neutralize international differences in
costs and hence competitiveness attributable to inter-country variations in environmental assimilative capacity or social preferences.

5.   While national sovereignty should be a governing principle for
efficient utilization of environmental assimilative capacity among
nations, export prices should refelect domestic environmental damages
and control costs. Thus, if through international negotiation a
furtherance of the provisions of the "polluter pays" principle as
set forth by the OECD countries is possible, we strongly recommend
that negotiation proceed as rapidly as is feasible.

6.2 Domestic Environmental Controls and Short-Term Trade Patterns

Two possible directions have been identified with respect to the short-
term impact of environmental management on trade patterns: measures
affecting production processes and hence unit costs, and measures

affecting the environmental attributes of the internationally traded
<u>products</u> themselves.

The most obvious source of possible international trade dislocations is regulation of the production process. If the "polluter pays"
principle--as accepted by the OECD member countries--is assumed, the
cost of reducing production-related emissions will be reflected in
final product costs and this will affect competitiveness. (Compare
chapter 7.1) If there are no other barriers to effective competition--
as within a unified national or regional market--the impact may be
rather pronounced. The international economy, of course, is characterized by a variety of other competitive distortions such as tariffs
and quotas, and hence the relative importance of short-term cost adjustments induced by environmental control may be diminished accordingly.

Apart from the existence of tariffs and other competitive distortions, the impact of environmental management on trade flows
generally and materials in particular will depend on two factors:
(1) the extent to which environmental control (EC) norms differ
between national states ; (2) the bearing of EC standards on production costs. When and if these standards begin to converge among
trading nations, the techniques employed in their implementation will
gain in relative significance, with the principal emerging issue being the use of public subsidies and assurance of trade neutrality. A
large number of organizations are already involved in standardization
which affect directly or indirectly environmental issues.* The most
promising are efforts to harmonize ways of <u>measuring</u> and <u>monitoring</u>
pollution, as opposed to the harmonization of environmental standards
themselves.

Within a nation, interregional variations in EC norms and
their competitive effects on trade are moderated by the existence
of a strong national government charged, in principle, with
minimizing their disruptive impact. Evidence in the U.S. suggests
that it will take a decade or more to devise a coherent system of
national environmental standards and enforcement procedures to attain this goal. Internationally, with political sovereignty much
more important and in the absence of supranational authority, the
adjustment period will be considerably longer and the competitive
implications for trade flows will be much more durable.

Within nations, due to the economic cost of interregional
variations in environmental <u>product standards</u>--combined with
asymmetries in power and interests of national vs. sub-national

---

* Examples include the International Standards Organization
(ISO), and the Food and Agriculture Organization (FAO), as well as
numerous international industry organizations.

political collectivities--a considerable degree of environmental
uniformity can be expected. Among nations, this conclusion does
not necessarily hold, (a) because the cost of differing product
standards to the world economy are not easily identified and
measured; (b) because such costs may be less pronounced given
large national or regional markets within which compatibility
exists; and (c) because differences in norms may be upheld by
virtue of national political sovereignty and in the absence of
supranational authority.

Nations will continue to translate their unique social
values and environmental conditions into product standards which,
they will (and should) insist, must be met by all goods and
services sold and disposed of in the national economy--whether
produced at home or abroad. Imports which do not meet these
standards will be barred from entry or discriminatorily taxed.
Whereas the range of products encountering this difficulty at
present is limited, it will widen substantially in the intermediate term. This raises three sets of problems.

First is the trade-restrictive impact of product-standard
variations themselves. Meeting such standards raises costs
even under a unified EC system.

Second is the problem of discrimination in the treatment
of domestically produced and imported products with respect to
their environmental characteristics. While inconsistencies in
standards may be disruptive of trade, they do not distort the
relative cost and price advantages upon which trade is based.
Indeed, they may reinforce them. However, when the standards
or their administration are applied to imports in a manner different from import-competing goods, they may take on the role
of a trade barrier.

Third, on the demand side it is clear that the cost and
quality changes brought about by product-related EC standards
will shift purchasing patterns. Consumers and industrial
purchasers will be induced to adopt less pollutive products
in place of those that are inherently more damaging to the
environment, and hence more costly to bring up to standard.
Then too, desirable characteristics of certain products may
deterioriate as a result of compliance with environmental norms--
examples of deterioration in automotive performance and the
quality of paints and pigments may be cited--forcing consumption
shifts within the affected product groups or between them.

Trade flows in certain products will thus be affected,
but this phenomenon will be intermittent and without a con-
.sistent pattern. The one area where a permanent demand shift
has affected trade involves furs, skins, and other products

from animals in danger of extinction (e.g., crocodile leather, seal skins, ivory, leopard skins, whale meat), as well as certain types of woods (e.g., redwood) subject to an extremely long growth cycle. All involve products entailing major environmental costs. These products have traditionally occupied an important but diminishing role in international trade, and can be expected to decline to insignificance in the intermediate term--with possible subsequent resurgence in specific products amenable to conservation and "controlled harvesting" practices.

Perhaps the most important sectors in international trade potentially affected by demand substitution are raw materials and fuels. The sulfur content of fuels has already produced dramatic shifts, and these can be expected to intensify. Trade in natural gas, an inherently clean fuel, can be expected to grow rapidly, particularly in Europe. And where pipelines are not a feasible mode of transport between origin and destination, trade in liquified natural gas (LNG) can be expected to develop. For the same reason, trade in nuclear fuels may also grow substantially in the longer run. There is also likely to be some shift in demand for raw material inputs from synthetics to natural products, provided that they are close substitutes and where the production of synthetics involves substantial environmental control costs. It may also increase the elasticity of demand for the products in question, thereby reducing their price volatility, although the problem of low supply elasticities remains.

There does not appear to be a complete study of lagged response or adaptive response to changes in export prices undertaken except for a few agricultural commodities. Based on the meager evidence available, a 30-70 percent adjustment to prices the first year with the remainder of adjustment spread over a 3-5 year interval might be expected. Thus, if a nation unilaterally imposes restrictive domestic environmental controls, one should expect fairly rapid adjustments in trade patterns as regards timing of response.

Recommendations

1.   Basic to any assessment of the short-term impact of environmental management on trade flows is a comprehensive inventory of national EC policies, plans, and implementational techniques. No such inventory exists. Appropriate international organizations should be encouraged to accelerate their work in this field.

2.   Substantial research is needed on the cost impact of environmental controls. Recent work on this subject in the United States, while preliminary in nature, provides a valuable start in this direction, and intensive research in this field must be encouraged as being basic to coherent policy formation.

3.   Careful empirical analyses of <u>international</u> demand relationships for basic materials is needed with emphasis on price responsiveness and timing. Evaluation is required of the extent of U.S.
influence on demand for basic materials internationally. Bureau
of Mines analyses in the United States provide an example of the
kind of systematic projections that must be undertaken on a global
basis.

4.   International product standards should be encouraged for those
goods that: (a) are traded in large volume internationally; (b)
have substantial environmental impacts; (c) where the environmental impact is clearly delineated and understood. Attempts should
be made to harmonize the adoption of product standards so that methods
and speed of implementation are compatible among countries.

5.   The question whether the incorporation of domestic EC costs into
traded products is politically realistic needs to be carefully investigated. This includes the problem of whether existing international institutions such as GATT are capable of encompassing EC
issues in their activities. Certainly this would be more desirable
than the establishment of new institutions.

6.   Since international product standards are likely to be developed only after implementation and testing of national EC norms,
it is imperative that nations be encouraged to hasten the unilateral adoption and implementation of domestic EC norms.

7.   The effect of product standards on international trade
patterns in the short run cannot now be assessed because of a
lack of information on product characteristics and standards,
adjustments in unit costs, or international demand relationships. It is likely, with such information gaps, that negotiation will need to proceed from those products with little
standardization required toward those where standardization
may be extremely difficult, except in those cases where non-
uniformity might lead to immediate disruption in trade patterns.

6.2.1 Special Considerations on Synthetic Organic Chemicals and Heavy Metals

The production of synthetic organic chemicals gives rise to residuals
which pose very difficult disposal problems. Furthermore, the consumption of plastics, synthetic fibers, and other synthetic organic
chemicals leads to a disposal problem. Some synthetic organics are
nearly indestructible and thus present litter problems, others de-
grade slowly, with unknown but potentially harmful effects on the
environment.

High disposal standards or taxes for residuals could cause a substantial part of synthetic organic manufacturing to relocate abroad. Imposition of consumption controls would impact on international trade, but would reduce the relative size of the production impact by reducing demand.

In the case of toxic metals (mercury, cadium, lead, chromium, and nickel), the bulk of U.S. consumption is now supplied from overseas sources, and even larger proportions will be so supplied in the future. The major environmental difficulty associated with these metals, so far as the U.S. is concerned, arises from their dispersal into the environment as a result of their intermediate consumption in a wide variety of industries and their use in end products.

Restrictions on the use and ultimate disposal of the toxic metals will probably reduce U.S. imports of such metals, although there may be offsetting adjustments for some heavy metals needed for pollution control equipment; i.e., cadium. The production adjustment burden on such restrictions will principally fall on the major supplier countries.

Recommendation

The emergence of special problems in the area of synthetic organic chemicals needs to be clearly recognized, including the development of possible trade barriers and adjustment difficulties for major foreign suppliers. To the extent practicable, efforts should be made to anticipate these kinds of problems and promote responsive adjustment policies on a multinational plane.

6.2.2 Domestic Environmental Controls, Domestic Costs, and Export Prices

According to a number of empirical studies analyzing production relationships for U.S. manufacturers, most industries, particularly export industries, seem to exhibit increasing returns to scale. This implies that the price adjustment in response to the imposition of EC costs will exceed the increase in costs when a new market equilibrium is achieved, and international trade effects may be magnified.

One can envision the possibility of two-part pricing systems not unlike agricultural commodities, or perhaps even tax rebate systems. Such systems should be discouraged as permanent offsetting forces for losses in sales to foreign markets. Rather, the principle of export pricing according to domestic social cost should generally be adhered to. Transitional export price adjustment problems might be handled through tax rebates if fixed time limits or other constraints were implemented on their use. Long-term loss in comparative advantage would best be handled by public policies aimed

at shifting resources toward areas of production with relatively
greater comparative advantage.

Estimated shifts in an index of U.S. export prices attributable
to domestic environmental controls anticipated for the future indicate a shift from one to four percent (Kneese and d'Arge, 1972).
This is assuming all control costs were passed on in the form of
higher prices of U.S. exports and import competing goods. Given
recent developments, it probably can be anticipated that the shift
might be substantially greater, especially for particular industries.

Recommendations

1. Studies of expected behavior by firms operating in an imperfectly
competitive industry in adjusting export prices in response to higher
domestic costs need to be undertaken.

2. Key industry studies are needed to forecast domestic costs and
export price shifts resulting from environmental controls. Also, estimates must be developed to gauge the loss in foreign sales and degree of domestic substitution.

3. Criteria must be developed for judging when an industry loses
long-run comparative advantage in terms of the environmental resources
available for its use and underlying its international competitive
position.

4. For some basic materials industries with relatively low elasticities of demand at the national market level and increasing returns to technologies of scale, the short-term impact on price and
revenues could be substantial. In consequence, it is recommended
that for these industries some flexibility be made in implementation
schedules for environmental control so that the short-run price
impact and loss in comparative advantage be attenuated. However,
such a policy should be of a short-term nature only with clearly
defined time limits and strict regulations on renewal.

6.2.3 Short-Term Balance of Payments and Income Effects

Improved environmental quality is not a free good. It must be paid for
by diversion of purchasing power either through taxes or through increased prices.

The international dimensions of the issue relate to two kinds
of effects. The first is a monetary question: If environmental
control is enforced through the general application of the "polluter
pays" principle, or if it is partly or wholly financed by government fiscal deficits, then it will result in a generally higher price
level than would otherwise be obtained. If the prevailing international financial system permits reasonable flexibility in exchange rates,

any such impact will be reflected in a commensurate depreciation of the national currency on the foreign exchange markets. If not, the result will be growing pressure on the current account of the balance of payments and, depending on what happens elsewhere, the eventual need for devaluation.

There are only a few econometric estimates of the impact of environmental control on the short-term balance of payments and level of aggregate income. One, using an econometric model of the U.S. economy and projected future American EC measures (assuming other nations do _not_ follow suit), emerges with the following results: Over the 1972-80 period, pollution control and compensatory macroeconomic policy measures are expected to raise the U.S. unemployment rate by 0.3 percent, raise the average annual rate of inflation by 0.26 percent, reduce fixed investment not related to pollution control by an average of $2.3 billion annually, and exert a negative impact on the U.S. balance of trade in the amount of $1.9 billion (Council on Environmental Quality, 1972). If, on the other hand, pollution control costs are 50 percent higher than current estimates, or standards are raised correspondingly, the average annual 1980 negative trade balance effect is estimated to be in the neighborhood of $3.2 billion or 6-7% of current level of exports.

Both unilateral and multilateral environmental impacts have been studied (d'Arge and Kneese, 1972). For the United States, with either unilateral or multilateral control programs, and with U.S. export prices increasing by 3.6 percent, there is a slight positive net change in domestic income. In both multilateral and unilateral cases the estimated net change was less than 3 percent: small enough to be considered nothing more than "noise" in the econometric model. Balance of payments shifts on current account amounted to a negative change of about 3 percent of 1968 exports in the unilateral case and a positive change of less than one percent in the multilateral case.

Recommendations

1. The meager empirical evidence available at this time suggests the balance of payments, employment, and national income effects resulting from domestic environmental controls are likely to be relatively small for the United States. This is true even if the trend toward indirect subsidization is reversed toward a full "polluter pays" principle. Thus, a tentative conclusion is that the _aggregate impact_ of domestic environmental controls should not be a matter of deep concern for policy makers. Of more crucial importance are the short-term impacts on particular firms and industries. Substantial additional research is needed on these questions.

2. Environmental control adds yet another reason for the development of a resilient international monetary system, one that embod-
ies sufficient exchange-rate flexibility to absorb the kind of impact likely to derive from this source. To the extent that such a
system emerges, the balance of payments implications of environmental management will be minimized.

6.2.4. <u>General and Special Adjustment Assistance Measures</u>

Special adjustment measures to assist those parties who are
driven out of jobs or businesses are embodied in the Trade Expansion Act of 1962, and it is tempting to recommend a counterpart
environmental bill. However, experience with special adjustment
programs reveals two basic difficulties:

> In any given industry or product line, there is always
> a distinction between more and less efficient firms.
> There is always the question whether the closure is due
> to inefficient management, new economic events, or
> other influences. The attempt to disentangle these
> factors by special investigations is a time-consuming
> process.

> 2. Assuming that impacted factories can be identified, it
> is not especially clear why those particular workers
> and capital owners should be assisted in meeting
> their adjustment crisis whereas other workers and own-
> ers who face an adjustment burden of equal magnitude,
> but arising from different economic circumstances, go
> unassisted.

These two difficulties provide a persuasive argument for the
superiority of general over special adjustment measures. In the
case of working men and women, assistance can be provided in the
form of unemployment compensation, retraining grants, early re-
tirement, payment of relocation expenses, and the like. In the
case of capital owners, compensation may be provided in the form
of more liberal carry-forward and carry-back loss provisions.

However, there are some persuasive arguments on the side of
special adjustment assistance as well. Clearly, those individuals
and industries adversely affected by domestic environmental control programs are not victims of natural market forces but of a
distinct shift in public preferences embodied in environmental
legislation. Domestic environmental controls can be viewed as not
unlike highway relocation or other public investments that remove
resources from individual citizens where these citizens are com-
pensated for their losses. In consequence, a special compensation
or retraining program might be applicable if competitive international disadvantages resulting from EC measures were substantial.

Recommendations

1.   The public as a whole benefits from improved environmental quality, therefore the public as a whole should help to defray the cost
of adjustment to the international competitive impacts of environmental management. We strongly recommend the formulation of a national adjustment assistance program designed to effectively ameliorate structural disturbances regardless of their origin. By increasing the mobility of resources in the national economy, the real
benefits of such a program will greatly outweigh its fiscal costs.

2.   This team strongly recommends that a study be made on general
versus special adjustment assistance measures for those adversely
affected by international trade impacts of EC. A corollary recommendation is to study the possibility of extension in tax loss recovery
provisions backward for those firms already or potentially affected.

6.3 Long-Term Effects of Environmental Policies on Trade Flows

It is difficult to estimate the impact of particular environmental policies on the exports and imports of commodity groups over
the long term. Trends in foreign demand and supply, technological
improvements, and foreign environmental policies are all difficult
if not impossible to forecast accurately. Moreover, a given policy implemented by the U.S. may cut both ways; that is, it may
discourage both imports and exports, with an indeterminate impact
on the net trade balance.

     Production and trade in particular commodities is often deter-
mined by rather small locational and know-how differentials and the
imposition of environmental controls can--in the long run--easily
offset these differentials with a large impact on the output of
particular commodities.

     Environmental management tends to divert a nation's productive
capabilities from things that can be traded internationally to
things that cannot. The question is whether or not the reduced potential output is symmetrical between those goods and services that
a country tends to import and those it tends to export. If it is,
the long-term EC impact is relatively neutral. But preliminary
evidence indicates that economic activities required for environmental management happen to absorb larger amounts of capital than
labor or other productive factors. If a country happens to export
capital intensive products and services while importing those
intensively using other productive factors, potential output of im-
portables (non-capital intensive goods and services) will decline
relatively less than potential output of exportables as a consequence of additional environmental controls. This may erode the
basis for a nation's comparative advantage in the international

marketplace. Hence the factor-intensity of environment control will effect the structure and volume of international trade, and each nation's gains from trade.

Even if environmental control standards are identical between nations, enforced identically, using the same mix of factor inputs, the impact of environment management may _still_ not fall evenly upon the trading nations of the world. The central issue is whether there are any systematic relationships between what a nation imports and exports and the degree of environmental damage associated with production of those particular goods and services. If pollution costs embodied in United States exports turned out to be higher than (or lower than) those of its imports, environmental control, even if uniformly applied by all of the major competitor countries, may have a disproportionate impact on the United States--either larger or smaller--along with other countries evidencing similar bias in over-all EC loadings on one direction or the other. The initial evidence is that EC costs per dollar of U.S. exports are about 15% higher than the EC costs per dollar of U.S. imports (Walter, 1973).

There is also the question of how _efficiently_ different countries go about implementing whatever EC standards they set for themselves. The fact that technology is highly mobile internationally would seem to suggest that EC innovations will in fact spread rapidly, and that the long-term implications of differences in technical know-how will be minor--and that trade in EC hardware and technology will itself become important, perhaps as an offset to negative EC trade-effects for capital intensive exporting countries in the forefront of environmental management.

Finally, international variations in EC norms and enforcement policies will affect patterns of industrial location according to (a) the importance of EC as a cost element, (b) the existence of differences in EC standards, (c) differences in labor availability and quality, public services, transport costs and proximity to markets, and (d) the nature of the firm's own logistical network, the character of its product, and its rate of growth in required productive capacity.

There is some evidence that such locational effects have already been felt interregionally within industrial economies. In the United States, for example, a number of political jurisdictions have begun to discourage industrial location and new investment, following an implicit cost-benefit assessment that the negative environmental con-sequences of such investment substantially outweigh the attendant static and dynamic economic gains. Particularly affected are the pollution intensive facilities. Firms that find perferred sites excluded for environmental reasons have available alternatives which encompass both domestic and foreign locations, and there are bound to be certain international locational spillovers attributable to local or regional environmental control.

To the extent that some domestic suppliers remain as import competitors, abiding by strict EC norms, governmental reaction to such locational flight into "pollution havens" may be expected. This may take the form of highly differentiated import controls as between basic raw materials and successively higher levels of processing, and block some of the gains expected from locational shifts. At the same time, serious questions will be raised about the social responsibility of business flight to pollution havens, with regard to its impact on the domestic economy, the environment in recipient countries, and global environmental balance. Virtually no conclusions can be reached thus far on the sensitivity of industrial location decisions to international differences in EC norms, except that the impact will tend to be longer-term in nature, often inseparable from other factors affecting such decisions.

Recommendations

1.   Insofar as is politically and economically feasible, shifts in long-term comparative advantage due to differing environmental control norms, or assimilative capacities should be allowed to evolve naturally. It is likely that natural long-term adjustments may reduce the volume of world trade, particularly for raw materials, as domestic expenditure is diverted to environmental quality improvement. This may not be the case for certain countries with comparative advantage in the export of pollution control equipment. Such a "natural" reduction in the volume and composition of trade among nations should not be viewed with alarm by policymaking bodies, except as it influences national security or other public goals.

2.   Too little is known about the kinds of productive resources absorbed by environmental control, or the implications of environmental management for national economic structures. Since these questions are basic for an assessment of long-term effects on the international trading system, we recommend that basic research efforts be pursued in these areas.

3.   Just as trade can promote the efficient use of environmental resources on a global scale, in accordance with assimilative capacity and social preference, so the same function can be served by international flows of productive resources. So long as there are no transnational environmental implications, such flows should not be impeded by "source" countries, and it is up to "host" countries to apply norms and procedures designed to enforce conformity with the level of environmental quality desired.

6.4 Special Issues on the Developing Countries

The developing countries do not have a common set of environmental problems nor one set of attitudes toward these programs. Some countries, such as India, Pakistan, and Brazil, have very substantial urban pollution problems; others, such as Indonesia and the Philippines, are seriously affected by the use of their tropical forest; still others, such as Egypt, are experiencing many environmental consequences--some foreseen and some not--of a substantial increment in the irrigation system. Again, environmental consciousness is higher in some countries, such as the Philippines and Singapore, than in others. Notwithstanding this diversity, both of problems and of attitudes, there is a tendency in the developing countries to assign a low priority to environmental concerns or to reject those concerns.

The existence of these negative views should serve to warn the United States and other developed countries that there is much diversity of opinion on the importance of particular environmental abuses, and the appropriate nature of corrective actions. Insofar as environmental abuses do not spill over national borders, the developed countries should be particularly sensitive to the preferences of the developing countries. And, in the case of DDT, polychlorinated biphenyls, and certain other toxic substances, where the use by one country adversely affects the ocean and air (common property resources of all nations) the developed countries should make every effort to secure the cooperation of the developing countries before imposing bans on the production or export of these materials.

Perhaps the most useful thing the U.S. and other developed countries can do at this stage is to augment the flow of knowledge on harmful substance and practices. It is important that developing countries set their own environmental policies; it is equally important that they set those policies with the maximum available information. Especially this is so in the case of substances and practices which will result in long-term, possibly irreversible, effects. Indeed, if the United States and other developed countries do not discharge their burden of supplying all available information to the developing countries, they will later be accused of withholding information, and reparations may even be demanded.

There is a widespread concern among the developing countries that international environmental measures in developed countries will have a negative economic impact on them. They fear that governmental expenditure of all types is relatively fixed in the developed countries and that environmental controls costs in these countries will come, at least partially, at the expense of aid funds for developing countries. Whether budgets are relatively fixed and whether developed countries will actually view domestic environmental control as a higher priority use of funds than foreign aid cannot be foreseen. Application of fiscal resources to EC needs may very well induce some net reductions in foreign aid--or reduce any net increases--but if the "polluter pays" principle or effluent taxes are adopted extensively, pollution abatement expenditures are not likely to have a serious impact on the level of foreign aid funds.

If the developed economies adopt simultaneously a domestic system of charges or standards for waste emissions control, their export prices will rise and profitability of investment in these countries relative to the developing economies will fall. Higher export prices would be expected to reduce exports of the developed countries and lower profit rates would divert international capital flows toward more profitable ventures in developing countries. Alternatively, extensive governmental financing of subsidies programs in developed economies is likely to have the opposite effect unless subsidies are tied explicitly to waste emissions damages. In the subsidies case profit rates will not decrease, but will more than likely increase. On balance, subsidies related to reductions in effluent discharges will have an impact on exports similar to, but probably smaller than, that of effluent charges but will have a different impact on international capital flows. Certainly, from the viewpoint of developing countries, emission standards or effluent charges adopted by the developed economies would appear to offer the fewest disadvantages in terms of both aid funds and trade repercussions.

Recommendations

1.   The United States and other developed nations should take the initiative in providing information on environmental degradation and alternative controls to developing countries. Such information should be freely given without implicit or explicit retaliatory implications.

2.   Explicit analyses should be developed on the impact of alternative control strategies, subsidies, standards, or taxes in developed nations on direct and indirect effects on import prices of basic materials.

3.   Aid to less developed countries is determined by a host of domestic and foreign needs and priorities, and a political assessment of the costs and benefits of such aid to the recipient and the donor nations. It is unlikely that the requisites of environmental management will materially alter these priorities to the detriment of the developing countries, and indeed we believe that the thrust of this kind of argument may be counterproductive to the interests of the developing nations. However, international discussions should be initiated as soon as practicable to assess the potential impact of environmental control on the developing countries, including the question of net resource flows.

4.   Several questions are still open; Will the probably lower environmental constraints on the developing countries be viewed as a natural
turn in the continuing evolution of international comparative advantage, or as basically unfair competition? Will the "unfair competition"
argument result in an implicit or explicit move to freeze the status
quo by means of protectionism on the part of countries importing materials produced in the "sweated environment"? These are basically
political and emotional issues which need to be thrashed out. We
recommend that the arguments on both sides be brought into the open
through appropriate research, but that no distortions of trade or
financial flows be imposed for this reason in the interim.

5.   We oppose the imposition of uniform environmental standards in
project evaluation on the part of international lending agencies'
operations in developing countries, but strongly support the use
of environmental impact statements designed to bring out the consequences of projects in this area. However, the small share of
international agencies in the supply of capital funds to developing countries may limit the importance of this procedure unless the
developing countries require such statements on their own. This is
to be encouraged.

6.5 Special Issues on Multinational Corporations

The multinational enterprise, by efficiently transmitting en-
trepreneurship, management, human skills and tangible capital in
a unified manner, can foster the rational exploitation of such environmental assimilative capacity in ways that are consistent with
optimal resource use from a global and perhaps from a national
perspective.

The environmental attributes of operations by multinational
firms within a given nation are dictated by policies formulated
at the corporate level which may or may not be consistent with the
environment in the host country--either in terms of its assimilative capacity or the social preferences of the national collecti-
vity. It is up to the host country to promulgate a coherent set
of environmental standards and administrative procedures that are
appropriate to its needs, and enforce uniform compliance by all
productive enterprises operating within its political frontiers.
The host country must recognize that the operations of multinational enterprises may be altered fundamentally by the imposition of
environmental rules, and by changes therein; that these may be
quite abrupt, and entail significant adjustment costs for the
national economy.

By operating within multiple national environments under a
single corporate structure, the multinational enterprise can potentially assemble, organize and centralize environmental know-

ledge for rapid application in its global operations, when required. Being subject to different standards in different countries at different times, it may be relatively versatile in its responses to policy shifts, possibly enjoying transitional competitive advantages in the process. Initial evidence indicates that executives of multinational enterprises already tend to standardize certain production processes on a global basis for reasons of intrafirm operating efficiency. Hence, although environmental standards related to productive processes may legitimately differ internationally, some multinational firms will behave as if harmonized standards existed.

Recommendations

1. Multinational firms should be treated as regards domestic controls, equally with domestic firms.

2. An international agreement should be developed on the availability of information pertaining to emission control of multinational corporations among nations currently or prospectively hosting such firms.

3. Multinational firms should be encouraged, where feasible, to apply less environmentally damaging production techniques throughout their global operations, since this can frequently be accomplished at only small increments in cost.

6.6 Aspects of Bulk Transport and Materials Supplies

Since the cost of transport is generally higher in international trade than in domestic trade, the foreign sector should be disproportionately sensitive to environmental control policies that affect the transportation industry. This sensitivity may take several forms:

1. Marine safety requirements. Apart from the general need for improved marine safety, a series of collisions involving tankships resulting in oil spills in the recent past have greatly increased pressures in this direction. Investment in anti-collision hardware and improved crew training, together with tighter enforcement of existing regulations have and will raise shipping costs--both directly and as a result of delays attributable to more cautious traffic control, particularly in congested areas.

2. Maximum size limitations. Economics in bulk transport are achieved primarily by increasing vessel size. But the risk of environmental damage per vessel probably grows with average vessel size, although not proportionately

because of the reduced number of ship movements required to
handle a given amount of cargo. Pressure is mounting to set
limits on maximum vessel size and, unless countered by advances in the safety area, may raise transport costs significantly above what they otherwise might be.

3.   Liability. Legal responsibility for accidental marine
      pollution is still being argued, but the trend is clearly
      toward increased liability for the ship owners, charterers,
      and shippers. This is raising marine insurance premiums
      and hence shipping costs. Full liability (or the less re-
      strictive OECD principle in this respect that the polluter
      does not pay residual damages, but only control costs) for
      environmental damage, however measured, could have a major
      impact.

4.   Operational changes. Environmental concern is also affect-
      ing standard shipboard operating procedures. Examples in-
      clude the flushing and cleaning of tanks at sea, sailing
      under ballast, disposal of shipboard sewage, and dumping of
      solid wastes. In each case the answer lies either in pro-
      vision of on-board treatment plants or in performing the
      necessary operations in port. Both are expensive in terms
      of capital requirements, as well as port charges, average
      running times, and other variable costs, and will raise the
      cost of sea transport.

5.   Port restrictions. Fear of environmental damage has closed
      several areas of development of ports, particularly for tank-
      ships, and most drastically on the U.S. East Coast. This will
      require sub-optimal rerouting of supply flows or the develop-
      ment of offshore facilities at higher cost, both of which will
      raise transport charges. At existing port facilities, strict-
      er anti-pollution requirements in cargo handling procedures
      will have a similar effect.

6.   Restrictions on transmission lines. The Alaska pipeline
      controversy is only the most visible of a growing array
      of actions against the environmental consequence of this
      form of transport. The disruption of domestic energy
      supply logistics will spill over onto the trade sector
      and lead to a significant impact on imports, exports, or
      both.

7.   Emissions regulations. There are relatively few air-
      pollution problems arising from marine and air transport.
      Particulate emissions from gas-turbine exhaust, apply-
      ing both to marine and aircraft propulsion units, can
      be remedied at relatively low cost, as can the noise
      problem in the case of in-port operations. Emissions

of particulates and other types of emissions from vessels
powered by steam turbines and diesels may be somewhat more
problematic, but involve the same kinds of approaches as
are required for internal and external combustion units in
land applications, both mobile and stationary.

8.   Differences among products shipped. It would appear that
      the pollution dangers deriving from accidental discharges
      at sea or in port vary significantly with the specific
      type of bulk material shipped. Clearly, the difference
      between accidental spillage of iron ore and crude oil is
      significant. But even within the fuels category, there
      is a major difference between crude oil and LNG; whereas
      the latter is inherently more dangerous to ship, it does
      not apparently pollute--although the capital costs in its
      transport are substantially higher. Consequently, the
      accidental discharge issue in bulk transport of materials
      seems to be highly product-specific, and to the extent
      that environmentally induced costs are important they will
      vary substantially by basic material.

      Transoceanic cost implications of environmental control for bulk
transport may be of importance as a determinant of physical trade volumes in the near term, but may be submerged by other forces operating on the supply and demand sides. In specific cases--e.g., block-
age of pipeline construction--and in the case of individual materials,
environmental control costs may have a serious impact on the existing pattern of international trade. There may also be induced substitutions between modes of transport (e.g., rail for sea or pipeline) and this may lead to a significant raising of costs. Finally, the economics of transport may have an effect of the in-place
cost of materials-recovery hardware, such as drilling rigs, storage tanks, and extractive machinery.

Recommendation

      Basic material policies should anticipate long-term changes
in transportation costs due to more stringent national and international environmental regulations on safety, maximum size, port
restrictions, emissions, and other facets of control. The impact
is likely to be more pronounced in the case of commodities for
which transport costs constitute a significant part of total delivered cost.

6.7 Recovery of Tradeable Materials

      A topic which may be of some importance in establishing links
between international trade and environmental controls is the recovery of resources that are internationally traded--or that are

substitutes for internationally traded products. This may be particularly appropriate in view of the stress in this report on resource recovery. The basic question is whether the materials-recovery loop--as it currently exists or is expected to develop--is basically intranational or transnational. Preliminary evidence is that it is transnational, involving major net importers and exporters of recovered materials.

As environmental standards are tightened, the range of reclaimed products widens, and since these products have to be taken out of the environment anyway, under prevailing EC norms, they may take on the characteristics of a "free" or very low cost input. Most of the kinds of presently reclaimed materials for which recovery prospects are good enter the channels of international trade. Hence one would expect that materials recovery will have a significant impact on trade flows, as a byproduct of domestic environmental protection.

Projecting this impact is difficult because it depends on (a) demand changes, (b) supply of virgin materials, (c) EC standards and enforcement, (d) economics of materials-recovery, and (e) logistics of materials recovery. Advances during the past decade indicate that large scale materials recovery from municipal wastes, sewage, industrial atmospheric emissions and liquid discharges, may become economic in the next decade. If parallel advances are made in collection and distribution systems, and if the economies of scale turn out to be favorable, then it can be expected that the impact of materials recovery on trade flows will be significant. Generally, countries with high levels of income and consumption will become important exporters of recycled industrial raw materials. The U.S. already exports metal scrap, rags, and paper in volume, and one would expect such exports to expand and to encompass a much wider array of materials, depending on international demand and supply patterns. Industries processing recovered materials, including chemicals, building materials, fertilizers, and pharmaceuticals, can also be expected to gain in competitive advantage in countries that have large flows of recyclables, as will industries that depend on such processed or unprocessed materials in the production of manufactures and semimanufactures. Even in the energy sector, economic incineration of nonrecoverable organic materials may conserve natural fuels and thereby affect trade flows. But again, direction and magnitude of the possible impact of materials recovery remains subject to a great deal of uncertainty.

Recommendations

1.   In order to promote the efficiency of resource recovery on a worldwide basis, every effort should be made to eliminate all barriers to international trade in recycled materials.

2.   Nations should endeavor to implement policies intended to promote substitution of recovered for virgin materials. Under certain circumstances, such objectives may be pursued by means of judicious application of export restraints, to avoid unnecessary depletion of nonrenewable resource stocks or imprudent exploitation of renewable materials endangering sustained yields.

3.   Considerable research remains to be done on new ways of using and storing recovered materials and their prospective role in international markets, as well as substitution between recoverable and virgin materials. Since nations differ greatly in their use of recycled materials, comparative research could yield substantial benefits.

6.8 Transnational Pollution, International Trade, and Basic Materials Policy

Transfrontier pollution is unique in terms of its impact on international trade patterns in that the initial effect is not transmitted through the price system but via biological and physical interdependencies among nations. In consonance with the recommendations adopted at the UN Conference at Stockholm, basic materials policy for each country should be designed so that it does not encourage residuals emissions into international common property resources. Second, basic materials policy should emphasize location decisions to reduce risks or hazards associated with use of common property such as the oceans and international estuaries, lakes, rivers, and common airsheds. Third, basic materials policy needs to be structured so that potentially hazardous residuals with global dispersion via food chains or other transportation processes are contained within the country of origin, or at least neutralized before crossing international boundaries. Fifth, basic materials policies should be shaped so as to encourage monitoring of emissions into transnational common property resources.

It is likely that gradual but extensive shifts in international trade and investment will take place in the area of renewable and nonrenewable resources as the relative scarcity of various materials changes over time. Major repercussions are already being felt in the fuels sector, discussed in detail in chapter 4. There are numerous policy questions that need to be answered in this respect. Closely related to environmental concerns are the following:

  1.   Should international trade controls and investment restraints be applied to the nonrenewables or renewables field? Such controls might be justified, for example, if it can be shown that the added demand of net exports and foreign investment in export-oriented resource exploitation results in unexceptable environmental damage or because of uncertainty may threaten to compromise prudent management of the affected resource.

2. Should international agencies be given a mandate for a
    program of extensive mapping and preliminary explora-
    tion of the earth's resources including environmental
    assimilative capacity so that the demand/supply puzzle
    might be seen in a better perspective?

3. What rationales, other than security of supply, can be
    advanced for the promotion of inefficient, high cost,
    U.S. domestic sources of materials as alternatives to
    lower cost foreign sources where a portion of the higher
    domestic cost is attributable to environmental control
    measures?

The policy issues raised here are stated in the form of questions
because the economic cause-and-effect relationships remain highly uncertain, as well as the role of noneconomic variables--such as geo-
political and strategic shifts in determining the appropriate direction of policy--discussed more fully in chapter 7.

Recommendation

International negotiations should be focused on transnational
problems. Once the list is drawn up, negotiations might usefully
proceed along the following lines:

1. A separate international commission might be established to deal
with each pollutant or identified abuse. Each member country would
tentatively agree to abide by the commission's decisions. A nation's
representation on the commission would depend both on its volume of
emission and on its volume of reception. Whether reception or emission was given greater weight would depend on the negotiations establishing the commission. Likewise, the plurality required for
a commission decision would depend upon these negotiations.

2. The commission would have the power to monitor national emission
and reception, publish data and reports, and, most importantly, to
levy taxes on offending member countries. The taxes would be paid by
the governments, and the governments might or might not choose to
deal with their internal problem using a taxing approach.

3. High priority should be accorded research on the global materials
balance, its impact on international trade and financial flows, and
the formation of policy on the part of both importing and exporting
countries. In the meantime, it should be recognized that countries
supplying renewable and nonrenewable materials and fuels have the
right--indeed the obligation to their peoples--to maximize the benefits they obtain from these resources over the long term, and that
the effort to achieve this objective may often lead to international
disagreements. The more soundly countries can plan, on the basis of hard
knowledge of the options and objectives of all sides, the fewer will be
the conflicts that emerge.

References

Brashares, J., 1971. "Cost Estimates for Environment Improvement Program" in Managing the Environment: International Cooperation for Pollution Control, A. V. Kneese, et al. (New York: Praeger), pp. 243-54

Council on Environmental Quality, Department of Commerce, and Environmental Protection Agency, 1972. The Economic Impact of Pollution Control. (Washington, D.C.: U.S. Government Printing Office), 332 pp.

d'Arge, R. C., 1971. "International Trade and Domestic Environmental Control: Some Empirical Estimates" in Managing the Environment: International Cooperation for Pollution Control, A. V. Kneese, et al. (New York: Praeger) pp.289-216

d'Arge, R. C., 1972. "Trade, Environmental Controls, and the Developing Economies" in Problems of Environmental Economics. (Paris: OECD), pp. 227-38

d'Arge, R. C., and A. V. Kneese, 1972. "Environmental Quality and International Trade," International Organization, vol. 26. n. 2, pp. 419-65

General Agreement on Tariffs and Trade (GATT), 1971. Industrial Pollution and International Trade. (Geneva: GATT), 35 pp.

Huynh, F., 1973. "The Relative Pollution Intensity of U.S. Trade". (Department of Economics, Michigan State University: unpublished), 30 pp.

Klotz, B. P., 1972. "The Trade Effects of Unilateral Pollution Standards" in Problems of Environmental Economics. (Paris: OECD), pp. 219-25

Koo, A. Y. C., 1971. "Environmental Repercussions and Trade Theory," Michigan State University, Econometric Workshop Paper No. 7102, 25 pp.

Leontief, W., 1970. "Environmental Repercussions and Economic Structure: An Input-Output Approach," The Review of Economics and Statistics, vol. 52, n. 3, pp. 262-70

Leontief, W., and W. F. Ford, 1972. "Air Pollution and Economic Structure: Empirical Results of Input-Output Computations," Harvard University (unpublished), 40 pp.

Magee, S. P., and W. F. Ford, 1972. "Environmental Pollution, the Terms of Trade and the Balance of Payments of the United States," Kyklos, vol. 25, Fasc. 1, pp. 101-18

Majocchi, A., 1972. "The Impact of Environmental Measures on Trade: Some Policy Issues" in Problems of Environmental Economics. (Paris: OECD), pp. 201-18

Ozorio de Almeida, M., et al., 1972. "Environment and Development: The Founex Report," International Conciliation, 84 pp.

Pearson, C., and W. Takacs, 1971. "International Economic Implications of Environmental Control and Pollution Abatement Programs" in United States International Economic Policy in an Interdependent World, Papers Submitted to the Commission on International Trade and Investment Policy. (Washington, D.C.: U.S. Government Printing Office) pp. 777-90

United Nations Conference on Trade and Development, 1972. The Implications of Environmental Measures for International Trade and Development. (Santiago: Document TD/130, UNCTAD), 20 pp.

Walter, I., 1972. Environmental Control and Consumer Protection: Emerging Forces in Multinational Corporate Operations. (Washington, D.C.: Center for Multinational Studies, Occasional Paper No. 2), 53 pp.

Walter, I., 1972. "Environmental Control and the Patterns of International Trade and Investment: An Emerging Policy Issue," Banca Nazionale del Lavoro Quart. Rev., n. 100, pp. 82-106

Walter, I., 1972. "Environmental Management and the International Economic Order," Ford Foundation-Brookings Institution Project on the Future of the International Economic Order. (Washington, D.C.: unpublished), 78 pp.

Walter, I., 1973. "The Pollution Content of American Trade," Western Economic Journal (forthcoming, March)

CHAPTER 7

STUDY TEAM

ON

INTERNATIONAL LEGAL DETERMINANTS OF NATIONAL MATERIALS POLICY

ABRAM CHAYES, Law School of Harvard University, Cambridge, Massachusetts, Chairman

HENRY BRODIE, Office of Environmental Affairs, U.S. Department of State, Washington, D.C.

MARSHALL I. GOLDMAN, Department of Economics, Wellesley College, Wellesley, Massachusetts

HANS H. LANDSBERG, Resources for the Future, Washington, D.C.

JAMES MacDONALD, Law School, University of Wisconsin, Madison

THOMAS F. MALONE, University of Connecticut, Storrs

PATRICIA RAMBACH, Office of International Environment Affairs, Sierra Club, New York, New York

PATRICIA L. ROSENFIELD, Department of Geography and Environmental Engineering, Johns Hopkins University, Baltimore, Maryland

RAYMOND VERNON, Harvard Business School, Cambridge, Massachusetts

RUBEN S. BROWN, Staff Officer

HENRY J. KELLERMANN, Committee for International Environmental Programs

7.1 Introduction

In its international aspects, United States materials policy over the coming decades will have to operate within a growing body of legal and quasi-legal constraints, designed for environmental protection. Domestic legislation itself has extraterritorial effects. Other countries, particularly in the developed sector of the world, are pursuing their own increasingly stringent environmental policies, sometimes alone and sometimes in concert. And the international community as a whole, notably at the United Nations Conference on the Human Environment in Stockholm during June, 1972, is coming to recognize that the human environment is a matter of international concern.

As a participant in, and at times as a founding member of, international programs and organizations dealing with environmental problems, the United States has accepted an increasing number of commitments which are bound to have an effect on its own national conduct, its policies, and its relations with other nations. Some of these commitments are limited in area of application, such as those stipulated by the "Guiding Principles Concerning the International Economic Aspects of Environmental Policies" developed by the Organization for Economic Cooperation and Development (OECD), by the terms of reference of the Committee on the Challenges of Modern Society (CCMS) of NATO, or the Economic Commission for Europe (ECE). Others have global character, such as the recommendations and resolutions of the Stockholm Conference or the International Conventions on Civil Liability for Oil Pollution Damage and Ocean Dumping.

The force of the constraints varies. Essentially, the effectiveness of the agreed on constraints depends on voluntary compliance within their own sovereign territory and elsewhere by the signatory nations, including the United States. In the case of the "Polluter-Pays Principle" adopted by OECD, largely upon the initiative of the United States, the reality of the United States commitment is beyond doubt. Its fulfillment, on the other hand, depends on the success of the United States Government in ensuring nation-wide acceptance of the principle by all appropriate levels of government and industry.

In the case of the 1972 Stockholm Conference, the situation is probably less unequivocal. Of the 109 recommendations some were adopted against the explicit opposition of the United States and other governments. But regardless of politics, none of the recommendations nor any of the principles in the Declaration of the conference are legally binding on any nation in any technical sense. Nor are they enforceable by the United Nations. Yet they create a political climate of expectations which is likely to influence profoundly the policies of many governments. National policies projected for a period extending over the next couple of decades must anticipate that these precepts will be translated into binding legal obligations at an accelerating rate.

The Stockholm Conference produced a series of principles which
in time may achieve the status of legal obligations regulating conditions within and relations between nations (United Nations General
Assembly, 1972). Of special relevance to the purpose of this report
are: Principle 1, which enunciates the fundamental right of man
to adequate conditions of life in an environment of quality; Principle
2 which postulates the safeguarding of natural resources; Principle
5 which stipulates the economic use and equitable sharing of nonrenewable
resources; Principle 7 which deals with pollution of the seas; Principle 14 which seeks to preclude conflict between economic development
and environmental protection; Principle 16 which calls for effective
policies to deal with problems of population density; Principle 20
which stipulates promotion of research and development to cope with
environmental problems; Principle 21 which establishes the responsibility
of governments for environmental damage inflicted upon other countries;
and Principle 22 which calls for the creation of international law
to fix liability and compensation for such damage.

The Conference developed an "Action Plan" which contained specific recommendations for environmental assessment, environmental management, and supporting measures. It prescribed specific courses of action
for the planning and management of human settlements; for the management
of natural resources, for the control of pollutants, for educational,
informational, social and cultural aspects, for development and environment. Specifically, it established principles which stipulated the
safeguarding of the earth's resources through rational management
and through an integrated approach to development planning. But of
overriding and critical importance was the emergence of the protection
and enhancement of the environment as a recognized priority for national
and international policies and of institutional arrangements to assure
continuity of world-wide governmental participation in the execution
of the "Action Plan."

United States adherence to international organizations, legal
agreements, and cooperative programs (whether multilateral or bilateral)
carries with it a dual obligation. On the one hand the United States
must examine decisions and actions on the international plane for
their applicability to the national scene. For example, Recommendation
59 of the Stockholm Conference called for a comprehensive study by
1975 on available energy sources, new technology, and consumption trends
in order to assist in providing a basis for the most effective development of the world's energy resources with due regard to the environmental effects of energy production and use. Current United States
initiatives to evaluate domestic energy policy might benefit considerably by close collaboration with this study to be conducted at
the request of the new United Nations environmental unit.

On the other hand, the United States must pay close attention to
the international implications of domestic policies and legislation.
Specifically, international environmental impacts of United States
materials policies may conveniently be considered in relation to three

groups of activities: (1) those carried out within the territorial
jurisdiction of the United States; (2) activities conducted in foreign
countries; and (3) activities in international environments, initially
and primarily the oceans. The last two categories will grow in importance with increasing reliance on external sources of supply, itself
partly a consequence of higher costs of domestic materials due to
environmental controls.

7.2 Activities within the United States

As already noted, United States production of resource commodities
and energy, like all other domestic industrial activities, is subject
to national and state environmental policies and legislation. These
are various and complex, but the principal national legislative vehicles
are the Solid Waste Disposal Act, the Resource Recovery Act, the Clean
Air Act, and the Federal Water Pollution Control Act, which more or
less regulate intentional and accidental releases of solid, liquid,
and gaseous wastes from industrial processes into the geochemical
environment and/or biota.

The evolving statutory framework also involves: (a) the pervasive
National Environmental Policy Act (NEPA) of 1969 which among other
things set up the Council on Environmental Quality, a White House level
policy center on environmental matters, and (b) several other statutes
dealing with external environmental effects not directly involving
the migration of substances (such as the Wilderness Act, the Noise
Pollution Control Act, and proposed legislation on land use, strip-
mining, sulfur use taxes, etc.).

Also important are the administrative and regulatory structures
(a) set up by the Executive Reorganization Plans No. 3 and 4 of 1970
which established the Environmental Protection Agency and the National
Oceanographic and Atmospheric Administration, and (b) the administrative
and evaluative framework still being set up throughout Federal agencies
to implement the environmental assessment provisions of the NEPA.

At present, United States antipollution standards in these fields
are probably more stringent and enforcement efforts more determined
than in most other countries. Despite this, we are, of course, one of
the world's chief polluters because of the high level of industrialization and resource use. At the same time, it does not appear that
we are presently in violation of operative international standards.

For activities having a direct impact on other countries, the
United States recognizes the principle of international law established
in the Trail Smelter arbitration that if activities within one country
cause measurable environmental damage to its neighbor, the first country
is liable to the second. The exact contours of this principle are by
no means clear, there is no reliable adjudicatory enforcement mechanism,
and in any case a damage remedy is much less satisfactory than preventive
action.

As a practical matter, issues of direct impact arise primarily
in relation to our two North American neighbors, Canada and Mexico.
In each case, a joint boundary commission has extensive responsibility
with respect to boundary waters. Like national regulation the activities of these boundary commissions affect resource production along
with all other kinds of industrial activity in the border areas.

An agreement was negotiated with Canada which gives the International
Joint Commission (IJC) a mandate to act as an independent monitor
of compliance with the agreement to achieve certain water quality
objectives for the boundary waters of the Great Lakes system. The
commission will not have positive regulatory authority, but its studies
and recommendations are likely to be very influential with national,
state, and provincial legislative bodies in both countries. Its work
on eutrophication of Lake Erie was influential in inducing restrictions
on phosphate detergents in communities on both shores of the Lake.

The International Boundary and Water Commission, United States
and Mexico, has been mandated to consider the problem of salinization
of the Colorado River due in part to irrigation and reclamation activities in Arizona. It should, and probably ultimately will, receive the
same kind of responsibility with respect to pollution of boundary waters
coextensive with IJC.

The IJC now has responsibility for international airsheds, but has
not been extensively employed for these purposes. The International
Boundary Commission, United States and Mexico, also deals mostly with
water problems but the 1972 bilateral environmental agreement between
the two countries provides an additional basis for action on airshed
problems. Both commissions should be vested with comprehensive
responsibility for trans-boundary environmental matters.

Some concern has been expressed in scientific and environmental
circles about the effects of intensive industrial activity on the chemical
composition of the global atmosphere (as opposed to local or regional
air quality) and on long-term climatic development. United States
industrial activity--and, from the standpoint of materials policy,
energy production--would represent a major component in any such
processes. In recent years well qualified study groups have examined
these questions (SCEP, 1970; RMOP, 1971; SMIC, 1971; SCOPE, 1972).
They have concluded, in general, that the existing state of knowledge
does not warrant drastic regulatory measures at this time. Each group
recommended continuing surveillance and scientific work looking toward
more accurate baseline measurements and assessment of the risks involved.
Recommendations in line with these judgments were adopted at Stockholm.
The United States should vigorously support such research and should of
course adapt its regulatory policy to the findings.

The principal international impact of domestic environmental
regulation in the United States is likely to be indirect--on the
cost and competitiveness of United States goods in international

trade and on the location of high pollution industrial processes
abroad. These issues are treated at length in Chapter 6. Some
members of our team expressed reservations about the effort to harmonize
national environmental controls on internationally traded goods as
among developed countries, which is the principal United States policy
response to this effort, and about the viability of an international
"polluter pays" principle. Many members of this team viewed the problem
of such cost differentials as a transitional one at least so far
as trade between developed countries is concerned and of little
consequence in the making of investment decisions. The OECD undertaking
represents a useful "holding operation," tending to forestall conflict,
and the inevitable "exemptions" will permit the more heavily involved
industries to live with it without destruction of the process of "convergence."

7.3 Activities Abroad

In general, this category comprises actions of United Stated nationals
or those financed in whole or in part by the United States government
In the materials field, this means primarily the exploration, development,
and exploitation of sources of raw materials, for the most part in the
developing countries. At present, much of this activity is in the private
sector, although there is some degree of governmental involvement in
almost every new project. Developing countries have maintained continuing pressure for "national sovereignty over natural resources,"
expressed in United Nations declarations and other forms. Trends in
economics and technological diffusion indicate that over the coming
decades local control over natural resources will be increasingly
asserted.

      Given this situation, the policy question is to what extent, if
any, the United States ought to project environmental standards to
the developing countries, either through regulation of its nationals
or as a condition of financing arrangements. Both the Stockholm
recommendations and the inherent characteristics of the issue dictate
a good deal of caution on this score. The absorptive capacity of the
local environment varies sharply from place to place, and is likely
in general to be greater in areas that have not been subject to heavy
and prolonged industrialization. Moreover, the balance between environmental and other values is necessarily a tradeoff that should be an
expression of local political processes rather than an imposition from
the outside. Some team members stressed the political and social
costs of United States indulgence of its penchant in recent years for
telling others what is good for them, and warned that the cycle should
not be repeated in the environmental field. At the same time, others
warned that failure to insist on uniform environmental controls now
might be criticized by future generations or governments in the developing countries. Chapter 6 stresses similar considerations in specifically evaluating trade and financial relationships with developing
countries.

On the other hand, although for purposes of analysis it may be possible to classify environmental impacts as local, transnational, and global, it is not so easy to make these distinctions in practice. And even the strongest proponents of local autonomy agree that it is strictly applicable only to local impacts. It may be that the effects of developing-country resource operations on the global environment, or on the environment in the United States itself, are rather limited. But it is probable that these activities will frequently have transnational effects in their own regions. In these cases, the national political processes of the country in which the project is located are likely to undervalue external effects; but other countries in the region may equally downgrade the economic values of the project to that country. There is no easy solution to this dilemma.

There is a strong case that, whatever the locus of the political process by which these competing values are weighed, the process should be an informed one. And this raises the question of the usefulness of environmental impact statements and similar devices. Should United States funding agencies be subject to the environmental impact statement requirements of the National Environmental Policy Act? The Agency for International Development (AID) and Overseas Private Investors Corporation (OPIC) have recognized this obligation as a matter of policy. The Export-Import Bank has not done so. Objections which may have been raised to the full-scale application of the impact statement requirements are not persuasive.

The impact statement requirement should be uniformly applicable to the financing and other activities of these agencies in the developing countries. But an impact statement filed in the United States does not meet the problem of alerting appropriate governmental or private agencies to the environmental considerations that may be at stake in the project. Thus, additional steps must be taken to bring the impact statement to the attention of host country officials so that they are given alternative choices. Host countries also might be encouraged to make their own impact statements. At the same time, care must be exercised to insure that so far as local impacts are concerned, this procedure does not acquire coercive overtones. The solution would seem to lie in fostering the capacity and desire of the host countries to engage in these activities and, at a minimum, to make the host country at least an equal partner in the exercise. However, the impact statement should not be used as a basis for withholding funds whenever the impact is wholly within the borrowing country.

The experience of A.I.D. might prove instructive. The agency has now developed procedures that require that information, technical data, analysis, and proposed solutions of environmental problems will be made available to the borrower/grantee and will accompany the project proposal through the review process with the borrower/grantee. A.I.D. views foreign assistance projects as not truly or exclusively United States projects and that the requirements of United States legislation; e.g., impact statements, as provided for in the National Environmental Policy Act, do not bind the borrower/grantee. A.I.D.

procedures stipulate, however, that the agency will work with the assisted foreign country or agency to ensure that project analysis and design reflect consideration of environmental factors and the alternative means, with associated cost/benefits, or minimizing undesirable environmental side effects while maximizing beneficial environmental results.

With respect to regional environmental impacts, there is no embracing political process to reconcile the competing views of the sovereign states affected. In this case, the outside investor or funding authority may have to take an important role in precipitating a solution. Here too, however, it should seek to apply criteria appropriate to the region involved rather than to export our own environmental ideas. When it comes to impacts on the global environment or on that of the United States itself, the imposition of whatever restrictions we think fit is justified, at least in principle. A difficult problem arises in the use of environmental regulation as barriers to trade, as by prohibiting the importation of items produced under less stringent environmental regulations than are applicable within the United States. If the purpose of the regulation is really to protect the health and safety of Americans there can be no objection provided the foreign-produced product is accorded the same treatment as the corresponding domestic product. But measures of this kind are apt to become vehicles for economic equalization, particularly as applied to materials imports, and would be objectionable as a new class of nontariff trade barriers. These questions are discussed in more detail in Chapter 6. Incidentally, several members of this team also concur with the recommendation in Chapter 6 on the importance of developing internationally agreed upon and locally recognizable warning symbols for materials which might be dangerous to human health and safety; e.g., the hundreds of mercury-related deaths recently in Iraq due, in part, to the fact that warnings on seed bags were written in a language not read widely in Iraq, reportedly Spanish.

Even "legitimate" import restrictions may seriously reduce earnings of particular developing countries from materials exports. Shifts to low sulfur fuels or low lead fuels and restrictions on furs, hides, and feathers of endangered species are examples of this phenomenon. As a result, developing countries have made demands for "reparations." At Stockholm a recommendation to this effect was passed over the objection of the United States and other developed countries. Given the relative positions of the parties, the developing countries are unlikely to be able to enforce these demands very soon, if at all. But materials policy must anticipate that this pressure will intensify and will focus on the United States and Japan, the major materials importers, whose environmental regulations will have a disproportionately large effect of this kind. It should be recognized that the two examples given above are analytically distinguishable. We are under no obligation to buy high sulfur fuels at the expense of our own environment, and no particular material supplier has a claim on the United States market. But the restriction on behalf of endangered species treats them as an international resource, and it does not seem equitable that the country of the habitat should bear the entire burden of their preservation. Thus, there is a stronger case in the latter situation for some kind of compensation.

The principal vehicle for United States private investment in materials production abroad is the multinational corporation. These entities have a strong tendency toward use of uniform technologies and thus to level up to the most protective standard of the countries in which they operate. Moreover, they are extremely vulnerable to political pressures from the host country on any matter that can be characterized as second class treatment. We should encourage multinational corporations to be a force for higher standards of environmental protection than would otherwise prevail in developing countries. It is sometimes suggested that codes of conduct governing foreign investment should be negotiated with developing countries. If so, environmental protection should be among the matters considered.

The United States also affects materials activities in the developing countries through its membership in international lending and financing agencies. The United States should press international lending and financing agencies of which it is a member to adopt, over time, policies adapted from the impact statement requirement of the National Environmental Policy Act. There is some support for such an approach in the Stockholm recommendations. The World Bank has developed detailed environmental assessment guidelines for its project evaluation and implementation procedures. Regional development banks have been less willing to adopt this approach.

Finally, the question of "additionality" was another issue that split developed and developing countries at Stockholm. Basically the question is how the extra costs of environmental protection measures in development projects should be borne. The developing countries argued that they should be paid for out of additional resources without invading sums made available for economic development by the international community. Again, the United States resisted on the grounds that the benefits of such measures accrued primarily to the host country. That conclusion is not necessarily sound. In the case of the establishment of a game preserve, for example, the community at large enjoys much of the benefit, and claims for additionality are strong. But even where the benefits are localized, the priorities of the host country and the discount rates at which long-term benefits are valued will usually differ sharply from ours. Pressures for some kind of international burden sharing in these cases are likely to mount over the decades ahead and ultimately to induce some kind of response from the donor countries and institutions.

7.4 International Environments: The Oceans

The oceans and the atmosphere above them constitute an environment largely beyond national jurisdictions and thus, beyond the responsibility of individual states for protection and regulation. At the same time, because of their relation to world climate and other features of the global environment, the oceans are an environment of particular importance.

They are important as well in connection with materials policy. The seabed and ultimately the sea itself will be increasingly relied on as sources of materials. And the oceans are important for the carriage of materials from foreign sources. Indeed, today petroleum products are the largest item in international seaborne commerce. Finally, the oceans have been the ultimate sink for the disposal of much of the waste from land-based production and consumption of materials, a function that may take on special significance with the shift to nuclear sources of energy.

All three of these functions involve serious hazards to the ocean environment. At a Law of the Sea Conference scheduled for November 1973, environmental protection issues will loom large on the agenda. There is thus a significant opportunity for United States policy to influence international developments.

As a matter of background, the oceans and seabed are divided for legal purposes into three principal zones: the territorial sea, the continental shelf, and the area seaward, which is wholly outside national jurisdictions. In general, though with some important qualifications, a country has complete sovereignty in its territorial sea, comparable to that exercised within its boundaries on land. In the years since World War II, most countries have recognized the right of the littoral state, first claimed by President Truman in 1946, to exclusive control over the exploitation of the resources of its continental shelf. Beyond the continental shelf, the oceans are governed by general international law, embodying essentially laissez-faire principles of free access and free exploitation for all. Recent years have seen a number of nations proclaiming limited regulatory authority in special functional zones seaward of territorial waters. Regulation of air approaches in the interests of national security was perhaps the first of these, but claims now exist with respect to fisheries and pollution regulation as well. The legal status of these claims is by no means clear.

If the scope of authority that may be exercised by a country within its territorial sea and on its continental shelf are reasonably well settled, the geographical scope of these two areas is at this time a matter of unresolved controversy. For a long time, the traditional width of the territorial sea was three nautical miles, but this consensus has broken down completely in recent years, with claims of up to 200 miles originating with states on the west coast of South America. The United States proposal for the 1973 Conference contemplates a

territorial sea of 12 nautical miles. Similarly, the seaward boundary
of the continental shelf is uncertain. It is defined in the 1958 treaty
on the continental shelf as 200 meters or the limit of technologically
feasible exploitation. It now appears that there is--or soon will be--
no such limit. So the United States is now proposing a fixed boundary
for the continental shelf at the 200-meter isobath.

The status of environmental regulation in these areas is equally
unsatisfactory. In the territorial sea, the action or inaction of
the national government with respect to pollution control is subject
to the same limited international constraints described above for
land-based activities. It appears that the requirements of United
States law will be a good deal more stringent than any international
limitations.

The Convention on the Continental Shelf contains a general antipollution principle, but no machinery for defining or enforcing the
scope of the obligation under it. And only 49 countries adhere to
the Treaty. Still it is a beginning of international law-making in
the area. In the United States, mineral exploitation of the continental shelf occurs under leases issued by the Department of the
Interior. New leases are presumably subject to the Environmental Policy
Act, and particularly the impact statement requirement. The Department
also prescribes operating regulations that do contain environmental
protection features.

Materials production will be heaviest in the inshore regions of
the seabed, that is to say, beneath the territorial sea and on the
continental shelf. These are also the most vulnerable areas from the
point of view of the marine environment, since as opposed to the deep
seas they provide a fragile, yet essential habitat for many forms
of marine life and are especially important to reproductive cycles.
From the point of view of both materials policy and environmental
concerns, there is reason not to press for the fastest possible
development of these sources. They constitute a reserve against depletion
of land-based sources. Improved technology that will be developed over
time will no doubt permit more effective exploitation. And both the
threats to the marine environment and the means of protecting it will
be better understood after some years, in view of the intensive work
now being done in the field. Research and development in these
issues should be strongly supported; and at the Law of the Sea Conference
the United States should work for strengthened international measures
for protection of the marine environment from resource activities on
the continental shelf.

Three major exercises are now under way for the development of
international legislation involving the protection of the marine
environment. They are (1) the Convention on the Prevention of Marine
Pollution by Dumping of Wastes and Other Matter; (2) the effort by
the Intergovernmental Maritime Consultative Organization (IMCO) to
eliminate pollution from the transportation of petroleum products;
and (3) the development of an international regime for the seabed
beyond national jurisdiction, with negotiations now going forward

in the United Nations Committee on the Sea Bed. All three of these
activities will impose important new international constraints on
materials policy. And although each is subject to criticisms in
detail, all three represent movement in the right direction from the
viewpoint of environmental protection.

The convention on ocean dumping is now open for signature. It
applies to the high seas, territorial waters, and bays. It contains
a list of toxic materials, dumping of which is prohibited absolutely.
There is also a "grey list," and a special permit from national authorities
is required for dumping of materials on it. Dumping of other materials
also requires a permit, to be issued by national authority subject
to criteria to be governed by certain considerations set out in an
annex to the convention. The convention as opened for signature governs
the disposal of nuclear waste. Annex I prohibits dumping of high-level
radioactive wastes and Annex II requires special permits for dumping
of medium- and low-level radioactive wastes. The dumping dealt with
by the convention is an important problem but not a major source of
marine pollution. It is important, therefore, that the effort at
adoption should not divert attention from more far-reaching measures.
Moreover, the convention is weak in its extensive reliance for implementation on national legislation and administration. Although it
provides for some service activities to be performed at the international
level, the institutional responsibility for these activities remains
to be specified. At least in the United States case, relevant national
legislation has also been passed, specifically the Marine Protection,
Research, and Sanctuaries Act of 1972.

IMCO has pledged the complete elimination by international agreement of pollution from ocean transport of oil. It has established
a target date of mid-decade, if possible, at least by the end of the
decade for certain. The United States should vigorously support
this effort, although it should be remembered that ocean transport
accounts for only about half of the oil pollution of the oceans; the
remainder derives from land-based sources. IMCO activity will probably
include a requirement that all tankers "load on top," which will bear
most heavily on tanker fleets of developing countries, further reducing
their competitive margins. This may cause particular problems for the
foreign exchange earnings of the traditional "flag of convenience"
countries--Panama, Liberia, and Honduras--and perhaps some others.
We have already moved to reduce those advantages in the area of taxation
and labor standards in response to our own view of the national interest.
We should not refuse now, in the face of an international environmental
claim. The question of regulating tanker size is discussed in the
chapter on international economic issues.

As regards the seabed beyond national jurisdiction, the United
States has presented a comprehensive proposal and draft treaty for
an International Seabed Authority. The primary purpose of the treaty
is to govern the development of the mineral resources of the seabed.

The draft agreement would divide the area into two zones. The inshore zone, corresponding roughly to the continental slope and rise, would be administered by the littoral state. Beyond that, the seabed would be administered by the international authority. The administering entity would issue licenses for exploration and production on the seabed, subject to the provisions of the treaty and rules, and recommended practices issued thereunder.

The draft treaty provides that "all activities in the International Seabed Area shall be conducted with strict and adequate safeguards for the protection...of the marine environment." And the Authority is empowered to make rules and recommend practices to implement this provision. Thus, as far as the delegation of power is concerned, the proposed treaty may be thought adequate. Since the proposed regime is primarily one for exploitation of the seabed, however, there is ground for concern whether the environmental protection provisions will be vigorously and effectively implemented.

The environmental aspects of the draft treaty can and should be strengthened by the addition of one or more of the following provisions:

1. the requirement of an environmental impact statement before any license is granted by the administering entity;

2. a requirement that each licensee make periodic reports analyzing in detail the environmental impact of the activities it has carried out;

3. a requirement that each license application contain detailed specifications as to the measure the applicant proposes to take to insure that his activities will be in compliance with the treaty requirement for "strict and adequate safeguards for the protection...of the marine environment";

4. that no licenses be issued until detailed rules and recommended practices for the implementation of this provision are promulgated in accordance with the procedures established by the treaty.

The proliferation of international texts and institutions has excited fears that both the basis and enforcement of legal obligations in the field of protection of the marine environment will become diffused and fragmented. Already there is support for an international agency with comprehensive responsibility for the control of marine pollution. It is likely that this support will grow over the decades ahead, and will find some expression in international action. On the other hand, existing agencies in the field and those that are likely to enter it soon tend to be tenacious and durable. They are unlikely to be displaced completely during this period. Separate functional agencies have at least a plausible claim that they are more effective than some centralized bureaucracy at formulating and operating pollution control programs relating to their normal activities. Still, it would seem desirable to make a start on the construction of a more comprehensive entity, and the Conference on the Law of the Sea is the best place for such initiative in the near future. In all probability, it would not be feasible to vest extensive regulatory authority in any such body at this stage. Its functions at the outset could be limited to information-gathering, inter-communication, catalytic and initiative functions, analogous to those vested in the environmental unit of the United Nations Secretariat mandated at Stockholm. Indeed, the preferred location for any overall marine environment protection responsibilities would be in the special section of the United Nations Environmental Secretariat. The United States should take the initiative in this direction at the forthcoming Conference.

7.5 International Environments: The Arctic

The Arctic Basin is an environment of great importance because of its fragility, its relation to global climatic change and because substantial reserves of natural resources yet unexploited appear to lie in the Far North. It is extremely vulnerable and the chief contemporary assaults upon it relate to the extraction and transport of its mineral resources. Canadian apprehension of damage to polar waters because of sea transport of North Slope petroleum through the Northwest Passage led at first to efforts at some bilateral regulatory arrangement with the United States. When this country was not sufficiently forthcoming, Canada, acting unilaterally, proclaimed a 100-mile wide zone in which she claimed jurisdiction to regulate ocean traffic for pollution control purposes. The United States has not recognized this claim.

On balance, the Canadian unilateral action is probably better than nothing at all. Nevertheless, the Arctic waters are an international environment, and it would be preferable to have other affected countries involved in their regulation. At a minimum, the United States should abandon its earlier apparent opposition and should actively seek far-ranging cooperative arrangements on these matters with the Canadians. Joint action need not be confined to the seas, but could extend as well to the protection of the tundra, overland transport of petroleum and other products, the ice-cap, research, etc. Such joint action could build upon prior scientific collaboration on climate, meteorological and biological conditions, and/or properties of the tundra and permafrost. Ideally, these arrangements should embrace all other countries surrounding the Arctic--the USSR, Finland, and the Scandinavian countries, in an Arctic basin organization. Here, too, it would be well for the United States to begin to move in this direction fairly promptly.

Appendix

Stockholm Conference
Actions Bearing on Materials Policy

I.   Declaration on the Human Environment

   A.   Proclamations

   4.   In the developing countries most of the environmental
        problems are caused by underdevelopment. Millions
        continue to live far below the minimum levels required
        for a decent human existence, deprived of adequate food
        and clothing, shelter and education, health and sanitation.
        Therefore, the developing countries must direct their
        efforts to development, bearing in mind their priorities
        and the need to safeguard and improve the environment.
        For the same purpose, the industrialized countries should
        make efforts to reduce the gap between themselves and
        the developing countries. In the industrialized countries,
        environmental problems are generally related to indus-
        trialization and technological development.

   6.   A point has been reached in history when we must shape
        our actions throughout the world with a more prudent
        care for their environmental consequences. Through
        ignorance or indifference we can do massive and irre-
        versible harm to the earthly environment on which our
        life and well-being depend. Conversely, through
        fuller knowledge and wiser action, we can achieve for
        ourselves and our posterity a better life in an environ-
        ment more in keeping with human needs and hopes. There
        are broad vistas for the enhancement of environmental
        quality and the creation of a good life. What is
        needed is an enthusiastic but calm state of mind and
        intense but orderly work. For the purpose of attaining
        freedom in the world of nature, man must use knowledge
        to build, in collaboration with nature, a better
        environment. To defend and improve the human environ-
        ment for present and future generations has become an
        imperative goal for mankind--a goal to be pursued
        together with, and in harmony with, the established and
        fundamental goals of peace and of world-wide economic
        and social development.

   7.   To achieve this environmental goal will demand the
        acceptance of responsibility by citizens and communities
        and by enterprises and institutions at every level,
        all sharing equitably in common efforts. Individuals
        in all walks of life as well as organizations in many

fields, by their values and the sum of their actions,
will shape the world environment of the future. Local
and national governments will bear the greatest burden
for large-scale environmental policy and action within
their jurisdictions. International co-operation is
also needed in order to raise resources to support the
developing countries in carrying out their responsibilities in this field. A growing class of environmental
problems, because they are regional or global in extent
or because they affect the common international realm,
will require extensive co-operation among nations and
action by international organizations in the common
interest. The Conference calls upon Governments and
peoples to exert common efforts for the preservation
and improvement of the human environment, for the benefit
of all the people and for their posterity.

B.   Principles

1.   Man has the fundamental right to freedom, equality and
     adequate conditions of life, in an environment of a
     quality that permits a life of dignity and well-being, and
     he bears a solemn responsibility to protect and improve
     the environment for present and future generations.
     In this respect, policies promoting or perpetuating
     apartheid, racial segregation, discrimination, colonial
     or other forms of oppression and foreign domination
     stand condemned and must be eliminated.

2.   The natural resources of the earth including the air,
     water, land, flora and fauna and especially representative
     samples of natural ecosystems must be safeguarded for
     the benefit of present and future generations through
     careful planning or management, as appropriate.

5.   The non-renewable resources of the earth must be employed
     in such a way as to guard against the danger of their
     future exhaustion and to ensure that benefits from
     such employment are shared by all mankind.

7.   States shall take all possible steps to prevent pollution
     of the seas by substances that are liable to create
     hazards to human health, to harm living resources and
     marine life, to damage amenities or to interfere with
     other legitimate uses of the sea.

13. In order to achieve a more rational management of
     resources and thus to improve the environment, States
     should adopt an integrated and co-ordinated approach
     to their development planning so as to ensure that
     development is compatible with the need to protect and
     improve the human environment for the benefit of their
     population.

14. Rational planning constitutes an essential tool for
reconciling any conflict between the needs of development and the need to protect and improve the environment.

16. Demographic policies which are without prejudice to
basic human rights and which are deemed appropriate
by governments concerned, should be applied in those
regions where the rate of population growth or excessive
population concentrations are likely to have adverse
effects on the environment or development, or where low
population density may prevent improvement of the human
environment and impede development.

21. States have, in accordance with the Charter of the United
Nations and the principles of international law, the
sovereign right to exploit their own resources pursuant
to their own environmental policies, and the responsibility to ensure that activities within their jurisdiction or control do not cause damage to the environment
of other States or of areas beyond the limits of national
jurisdiction.

22. States shall co-operate to develop further the international law regarding liability and compensation for
the victims of pollution and other environmental damage
caused by activities within the jurisdiction or control
of such States to areas beyond their jurisdiction.

References

Committee for International Environmental Programs (IEPC), 1972.
    Institutional Arrangements for International Environmental
    Cooperation. (Washington, D. C.: National Academy of Sciences).
    74 pp.

McNamara, R. S., 1972. Address to the United Nations Conference on
    the Human Environment. (Washington, D. C.: International Bank
    for Reconstruction and Development), 14 pp.

NASA Langley Research Center, 1971. Remote Measurement of Pollutants
    (RMOP), Report of a working Group, Norfolk, August 16, 1971.
    (Washington, D. C.: National Aeronautics and Space Administration)

Scientific Committee on Problems of the Environment (SCOPE), 1972.
    Global Environmental Monitoring, a report submitted to the United
    Nations Conference on the Human Environment, Stockholm, 1972.
    (London: SCOPE), 67 pp.

Study on Critical Environmental Problems (SCEP), 1970. Man's Impact
    on the Global Environment. (Cambridge: Massachusetts Institute
    of Technology Press), 319 pp.

Study of Man's Impact on Climate (SMIC), 1971. Inadvertent Climate
    Modification. (Cambridge: Massachusetts Institute of Technology
    Press), 308 pp.

Swedish Royal Ministry for Foreign Affairs and Royal Ministry of
    Agriculture, 1971. Air Pollution Across National Boundaries:
    The Impact on the Environment of Sulfur in Air and Precipitation,
    a report by the Swedish Prepatory Committee for the United Nations
    Conference on the Human Environment. (Stockholm, Sweden: United
    Nations Conference on the Human Environment Secretariat), 96 pp.

United Nations General Assembly, 1972. Report of the United Nations
    Conference on the Human Environment held at Stockholm, 5-6 June
    1972. (Geneva: A/Conf 48/14, United Nations Conference on the
    Human Environment Secretariat), 124 pp.

CHAPTER 8

ENVIRONMENTAL AND RESOURCE PROBLEMS AND POLICIES
IN JAPAN: A COMPARATIVE CASE STUDY

CHARLES S. DENNISON, New York, New York

The situation that the United States faces in managing the environmental aspects of a materials policy resembles that of other industrialized countries. As Japan is the world's third largest industrial producer and its largest importer of raw materials, the committee asked one of its members to prepare a synoptic report on the Japanese situation while on a visit to Japan in September 1972.*

While Japan was selected for this case review, the committee recognizes that similar environment-resources situations exist in the other advanced societies--the nations of the European Community, the USSR, and Canada--and that very different but most urgent problems are encountered in the developing countries.

This summary of the report describes the Japanese environmental condition briefly and then selects several major Japanese programs and developments for comment. Certain aspects of the international side of Japan's environmental-materials problems are discussed, concluding with recommendations for United States action.

8.2 <u>Economic Advance and Environmental Costs</u>

Japan's phenomenal post-war economic advance has brought serious, relatively sudden environmental crises and problems, unsupportable industrial and population concentration and a range of materials and resources problems. The severity of this situation is caused by the nation's driving economic development and growth--trade surplus in 1972 was $8,972,000,000--pressing against the limits of its crowded, resource-scarce land mass.

Japan's commitment to economic growth as an expression of national will to place and power in the world will evidently continue in at least the coming decade or two. The thrust to catch up with or surpass the West confronts tightening environmental limitations, rising costs, and the public and political determination to force remedy, nationally and locally. The collision of these forces will undoubtedly be accomodated within the unique Japanese system of governance, a compact hierarchical administration from national executive to prefecture, interlocking power groups, working through a process of exhaustive prior assessment and consultation, which usually leads to consensus decision and disciplined implemental action. Unhampered by the delay and attrition of adversary action characteristic of western societies, the Japanese system could produce synergistic action that will permit a modified, redirected, probably curtailed, but competitive economic advance while protecting, even restoring the environment.

---

*With the cooperation of the U.S. Embassy in Tokyo and helpful assistance from Japanese ministries, agencies, and private industry.

8.3 Basic Policy--Prime Minister Tanaka's Program for the Remodelling of the Japanese Islands

This drastic program which the Prime Minister calls the "path for a New Japan," is "The foremost challenge of my domestic administration and... is anticipated to influence the economic relations Japan has with other nations to a considerable degree." Mr. Tanaka explains the rationale for the program:

> "All the political, economic, social, cultural and environmental problems for all human beings sharing a common destiny on this Earth are approaching a drastic turning point. A new order and new orientation are called for in every arena and sphere."

The Prime Minister explains his purpose:

> "The implementation of this project is expected to drastically solve the problems of mammoth cities, rural depopulation, environment, prices and mental decadence, and to help direct the industrial potential of Japan which has come to occupy a notable position in the world economy, from the production or export of private consumer goods to the supply of public goods."

Expanding on this theme, Mr. Tanaka proposes to shift the thrust of the dynamic Japanese economy from "pursuit of growth" to "use of growth", a long-term, measured, comprehensive process in the course of which he believes that "a new frontier other than private consumption, private capital investment, or export may emerge for industries."

The remodelling of the Japanese Islands is consistent with the nation's comprehensive plan adopted in March 1969, and running through 1985. Its principal objectives are:

The relocation of industries by movement from congested areas and the deliberate construction of plants in sparsely populated regions to be achieved by fiscal pressure and incentive and the provision of infrastructure and services. Environmental conservation and land utilization will be strongly enforced.

The establishment of high-speed transportation and information networks to change the nation's economic and population flow. This comprehensive system based on a massive high-speed public transport grid, already represented by the "bullet train" lines, will be integrated with highways, harbors and pipelines. The complementary communications network will employ a real time data transmission system, television-telephones, etc., in a rapid functional instrument for programmed nation-wide development.

The building of new local cities designed expressly to attract big
city residents. These new cities of about 250,000 population will
be strategically located on the transportation network, each offering an appropriate mix of services, industrial parks and leisure
facilities.

Prime Minister Tanaka's remodelling program is political and con-
troversial and has many critics. For example, opposition parties charge
it has given rise to extensive land speculation. As in all democratic
societies, many conflicting views and interests are involved. Delays,
obstruction and changes must be expected as the Japanese system works
its way. This is, however, a bold statement by a newly elected premier
who has grasped the meaning of the times and sought to direct his country
toward prudent action.

8.4 Specific Environmental Problems

Japan is grappling with an acute environmental situation--its problems
are visible, harmful, and politically sensitive. The nation is small
physically, 370,000 km^2, of which 106,000 km^2 is populated, GNP per km^2
(in US$) is 1860 vs. 268 for the U.S. Despite the wide difference in
land area, the U.S. does have massive population concentrations--over
60% of its population lies in coastal and inland megalopoles where accumulated environmental problems and impacts are similar.

Japan's policy of dynamic industrialization in the post-war era,
chiefly in the energy-based heavy and chemical industries, caused sharp
increases in importation of raw materials from abroad and their processing and consumption. Energy consumption tripled from 1954 to 1964, and
doubled in the next five years. (However, per capita consumption of
energy in Japan is still far less than in the United States.) Motor
vehicle population has jumped four to five times in the past ten years
and at 20,000,000 vehicles, it is second only to the United States.
These increases in material throughput and consumption have caused
alarming rises in pollution and environment degradation.

8.5 Environmental Control

A series of laws for controlling pollution and environmental degradation
have been enacted starting with the Water Quality Conservation Law in
1958, the Regulation of Soot and Smoke in 1962, and reinforced with the
1967 Basic Law for Environmental Pollution Control. The more recent law
concerning Entrepreneurs Bearing of the Cost of Public Pollution Control
has established the polluter pays principle. A new law fixing absolute
liability for polluters went into effect October 1, 1972.

The courts are actively enforcing laws acting on behalf of citizens
harmed by environmental excesses. Signal examples are the Nagoya High
Court's upholding of a lower court decision that the Mitsui Mining and
Smelting Company was responsible for causing the "Itai-Itai" disease

(cadmium poisoning). Other actions involving "Minamata" disease (mercury poisoning) and SO_2 emissions in Central Japan indicate sharpening judicial action against industrial polluters.

8.6 Efficiency and Conservation in Materials Use

Government policy arrived at characteristically in concert with industry encourages the development of technology to obtain better resources utilization from both aspects, conservation and environment. Programs involve reduction of unit consumption from improved processing methods, substitution of materials, use of lower grade ores, exploitation of unrecovered resources, and a recent program for waste utilization, materials recovery and recycling. In this latter field Japan seems to be well behind the United States in both technology and programming.

The Smoke and Soot Regulation Law of 1962 and subsequent legislation has brought about marked technological advance in treatment of stack gasses resulting in recovery of sulfur sufficient for national needs.

8.7 Government Organization

The Japanese government apparatus for management of environmental and resource matters was evolved over the past 20-25 years. It reflects the unique hierarchical system of interlocking governmental and private power groups, of thorough advance study, appraisal and consultation, usually leading to consensus decision and disciplined action.

The center of authority is the Prime Minister's office (as shown in the accompanying chart). The Economic Planning Agency has overall responsibility for long-range economic land use and water utilization policy.

The Environment Agency established in July 1971 is charged with setting and enforcing quality and emission standards.

Within the cabinet, various ministries have specific fields of responsibility as indicated on the chart. The most important by far is the Ministry of International Trade and Industry (MITI) that is charged with the actual execution and planning of a range of vital environmental resource activities.

MITI is the agency charged with guiding, encouraging or restraining private industry at home and abroad. It employs a well-tested set of incentives and disincentives to guide the private sector on desired national issues. It can assist in provision of capital and through special depreciation write offs to hasten modernization of plants. One fundamental reason for expecting Japan to effectively harness her national economic interest with environmental management is the government's--and largely MITI's--clear record of success in shifting and modernizing Japanese industry in the 50's and even more so in the 60's. This capacity to foster and achieve selective modernization while ruthlessly eliminating obsolete and inefficient enterprises should be useful in the growth _vs._ environment adjustment.

All ministries and agencies rely on advisory councils or expert committees of mixed governmental, academic, and private industrial membership for advice and recommendation.

Chart 8.1:   Government of Japan Agencies with Natural Resources and Environmental Affairs Responsibilities

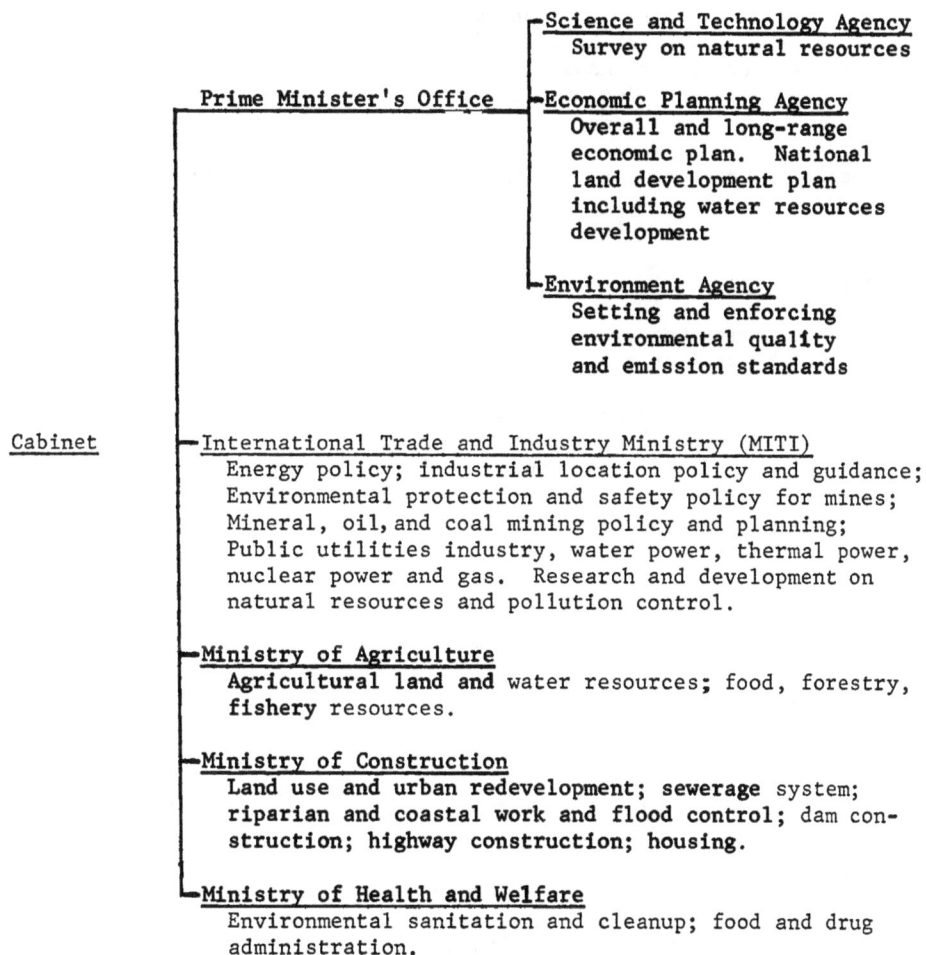

Cabinet

— **Prime Minister's Office**

┌─**Science and Technology Agency**
 Survey on natural resources

├─**Economic Planning Agency**
 Overall and long-range economic plan. National land development plan including water resources development

└─**Environment Agency**
 Setting and enforcing environmental quality and emission standards

— International Trade and Industry Ministry (MITI)
 Energy policy; industrial location policy and guidance; Environmental protection and safety policy for mines; Mineral, oil, and coal mining policy and planning; Public utilities industry, water power, thermal power, nuclear power and gas. Research and development on natural resources and pollution control.

— **Ministry of Agriculture**
 Agricultural land and water resources; food, forestry, **fishery** resources.

— **Ministry of Construction**
 Land use and urban redevelopment; sewerage system; **riparian and coastal work and flood control;** dam con- **struction; highway construction; housing.**

— **Ministry of Health and Welfare**
 Environmental sanitation and cleanup; food and drug administration.

8.8 International Resources Implications

Continued rapid Japanese industrial expansion is significantly related
to the world's sum availability of natural resources, chiefly minerals.
The trade, investment, balance of payments, and political implications
of competition between the two economies in raw materials acquisition
should sharpen as the United States moves abroad for an increasing share
of its requirements. Common environmental pressures should cause both
countries to compete for certain materials, particularly low-sulfur
crude oil and liquid natural gas. On the other hand, the huge capital
requirements and the scale of prospective minerals ventures are encouraging joint United States-Japanese ventures. Such multinational ventures involving participation of consumer nation governments (e.g., the
United States and Japan), producer nation governments, and local and
foreign private enterprise should offer vehicles for mobilizing capital
and technology while ensuring an equitable and stable international raw
materials market. (See appendix for excerpts from the White Paper.)

Japan's approach to its energy problem illustrates many of these
implications.

8.9 Energy Strategy

Japan's Comprehensive Energy Council has drafted a specific strategy in
this field which has four aspects:

1. Reduction of sulfur content of crude oil used and installation of desulfurization capacity. This follows the
   government's "Sulfur Content Reduction Program," MITI.

2. Use of liquid natural gas

3. Nuclear power

4. Improved energy technology

8.9.1 Sulfur Reduction

1. Crude Supplies

   The switch to low sulfur fuel has had and will continue to
   have marked effect on Japan's procurement practice (the
   average sulfur content of imported crude oil decreased from
   2.04% in 1965 to 1.68% in 1969 and a goal of 1.5% is set
   for 1973). In particular, it bears on the country's purchase
   of low sulfur crude from Indonesia which accounted for 71%
   of Indonesia's exports in 1970 and 73.5% in 1971. In May
   1972 a government-to-government agreement provided a con-
   cessionary loan by the Government of Japan of US $207,000,000
   for petroleum development in return for the supply by Indonesia
   of 50 million metric tons of low sulfur crude oil over the
   next ten years (in excess of supply through normal commercial
   channels).

Japan is also seeking sources of low sulfur crude throughout the world. A keen awareness, heightened by OPEC actions, of her vulnerability to supply curtailment makes dispersion of purchase (including China, Russia, and West Africa) a prime factor in contract arrangements.

The nation's supply vulnerability has caused increased exploration, particularly offshore the Japanese archipelago where the limited supplies developed have shown a low sulfur content (0.1%).

2. Desulfurization

Government standards and public pressure have caused a major effort toward sulfur recovery from stack gas and from crude oil. At the end of 1971, the crude desulfurization capacity was 510,000 barrels per day and the Petroleum Association of Japan has estimated increases in capacity to 755,000 b/d in 1972; 911,000 b/d in 1973; 1,018,000 b/d in 1974; and 1,163,000 b/d in 1975.

8.9.2 Liquid Natural Gas

Japan's search for nonpolluting fuels has led to aggressive development of liquid natural gas sources abroad. Typical is the Tokyo Gas Company-Phillips-Marathon venture in Alaska which will provide 720,000 tons of LNG per year. A long-range venture with Shell in Brunei, North Borneo, is to provide 3.5 million tons of LNG per year for 20 years. There is a similar venture in Iran and a prospective venture in the USSR (Siberia), China, and other source countries. Of particular interest is the Tokyo Gas Company-El Paso Natural Gas Company venture in Yakustian, Siberia, announced on October 29, 1972, involving the possible export of 1,050 billion cubic feet of gas per year to be divided about evenly between the United States and Japan.

8.9.3 Nuclear Power

About 1965, in the face of deep nuclear aversion, Japan embarked on a long-range nuclear power program. Four major installations are in operation and a number of 1100 megawatt units are planned--the suppliers being Toshiba, Hitachi, Mitsubishi Heavy Industries, and General Electric and Westinghouse. The government and industry are following breeder reactor developments throughout the world and are conducting their own intensive research and development at a new experimental fast breeder reactor at Orai. Japan contributed a million dollars a year for six years to the Detroit Edison breeder reactor development at Monroe, Michigan. The country's growing nuclear needs have raised the question

of supplies of enriched uranium. The United States enrichment plants
will not be adequate for United States and Japanese total needs. As a
result the electric power industry has entered talks with Reynolds
Metals Company, among others, on a possible joint venture, with capital
investment of $2.5 billion, to produce enriched uranium in the United
States. The plant would produce enriched uranium of 8,705 ton SWU's
annually and would commence operation by 1975. Despite progress with
nuclear power there is very stiff opposition to siting of nuclear power
plants by local citizenry who fear a radiation threat.

8.9.4 Improved Energy Technology

MITI policy states that, from the long-range view, it is essential to
reduce the country's consumption of resources while maintaining at
least the economic results equivalent to those that would have been
gained without decreasing the use of resources. Other than shifting
from heavy to knowledge-intensive industries, the answer lies in developing better resource utilization techniques and in technology for
tapping or producing new types of resources that will be available in
almost unlimited quantities.

8.10 Position in the International Raw Materials Situation

Japan's growing dependence--now 90% and expected to reach 93% by 1975--
on imported raw materials has caused her Ministry of Trade and Industry
to make a comprehensive analysis of present and long-range prospects.
(See appendix for excerpts from the White Paper.) The MITI paper notes
that, "The international situation on the supply of...resources does
not allow of optimism. Progress in material civilization induced by
the universal desire to raise the living standard has certainly increased per capita consumption of resources. This, coupled with the
explosive growth of the world's population, has produced a multiplying
effect on the consumption of resources. In the 1970's consumption is
expected to reach enormous proportions...."

"However the world's production and supply of these resources is
beset with growing uncertainties in a fluid situation, in which developing countries that produce resources are making nationalistic demands,
advanced industrialized nations are moving to secure resources at low
cost and on a stable basis, and major international interests are trying to strengthen their unity in the wake of threatened collapse of
their oligopolistic organization."

Faced with this prospect and with the exacerbations caused by environmental problems, Japan is following a course that:

1.   plays a direct and aggressive role in the international
     resources economy;

2. reduces her reliance on simple importation through foreign,
    including United States, international corporations;

3. recognizes the thrust of developing countries toward devel-
    opment by encouraging and financing their indigenous pro-
    duction and upgrading of materials;

4. uses her colossal purchasing power in contract relations to
    maintain favorable, stable sources of supply; and

5. stipulates that "technology designed for economies in re-
    sources should be developed while the industrial structure
    should be re-oriented with priority given to knowledge-
    intensive industries.

8.11 Conclusions and Recommendations

8.11.1 Conclusions

The United States and Japan, as the world's leading free economies, face
many of the same environmental and materials problems as they enter the
final quarter of the century. While the United States has more severe
overall problems because of its level of technological advance and the
degree and scale of its industrialization, Japan faces perhaps more intensive pressures because of her restricted land mass.

Both countries are entering the post-industrial phase of their
history--a phase in which services and the development and use of knowl-
gradually exceed the production of goods as the main concern of the
economy.

Both countries have recognized that the health and the quality of
life of their citizens are of prime national concern and must be advanced as a matter of national policy.

Both nations--particularly the United States--have the capability
for measuring and anticipating change and its impacts, and both can use
this capability to try to shape the future as a matter of deliberate
choice.

Both nations--with the United States well in advance--have evolving national legal structures and governmental institutions designed to
grapple with rising environmental problems.

Each nation has much to gain from the other, even as each nation
competes with the other in the world resources economy.

8.11.2 Recommendations

That current consultative procedures between both governments be continued and extended at all levels--executive, foreign policy, economic
policy, materials policy, environmental policy, and on specific technical problems.

That both governments implement effectively the Guiding Principles
Concerning the International Economic Aspects of Environmental Policies,
adopted by the Organization for Economic Cooperation and Development
(OECD) in May 1972. Among other topics, these Principles deal with cost
allocation (the "Polluter-Pays" principle, harmonization of environmental standards, and the need for international consultations between
the OECD governments concerned.

That as the United States evolves its own materials-energy policy,
practice and organization, it take into account Japanese experience
and adjust the consultative process to meet new governmental patterns.

That the present consultative and collaborative procedures between
the private sectors, academic and industrial of both countries be
continued and extended.

That both countries give close consideration to the implications
of growing competition between them for supplies of certain scarce and
environmentally desirable raw materials and that both work toward cooperative or collaborative development of such resources while supporting an open nondiscriminatory trade system.

That both governments recognize that the scale, complexity, sensitivity, and cost of major minerals development projects suggest the need
for multinational ventures involving consuming and producing country
governments as well as private enterprises.

That both countries seek to harmonize their relations with raw
materials-producing developing countries, particularly in matters of
development and technical assistance, thus avoiding the tensions and
risks of power rivalry.

That both countries use their world influence to support the international organization and programs proposed at the United Nations
Conference on the Human Environment, Stockholm.

The recommendations made above also apply to United States bilateral
relations with other industrial and developing countries.

APPENDIX

Japan's Natural Resources Situation

(excerpts from and comments on the
White Paper on Natural Resources
Problems in Japan, issued by the
Ministry of International Trade
and Industry (MITI), Tokyo, 1972)

Environmental Impact on Natural Resources

The Japanese economy has been expanding at an unprecedented tempo
until recently. This remarkable economic growth has been supported
by massive imports of natural resources. Confined within a corner
of the Orient in the wake of World War II, Japan's economy expanded
gradually in the postwar era, buying large quantities of resources
from practically all parts of the world to turn out a wide array
of products from its giant economic mill.

At the beginning of the 1970's the nation's resources imports
reached the largest scale in the world. Japan also ranked first
in consumption of major resources per unit of GNP. However, the
international situation on the supply of these resources does not
allow of optimism.

Progress in material civilization induced by the universal
desire to raise the living standard has certainly increased per
capita consumption of resources. This, coupled with the explosive
growth of the world's population, has produced a multiplying effect
on the consumption of resources. In the 1970's consumption is
expected to reach enormous proportions. For example, cumulative
consumption of petroleum in this decade is estimated at a level
equal to the total consumption of this key resource for the past
110 years. This is symbolic of the rapid tempo at which the
world's consumption of natural resources is expanding.

However, the world's production and supply of these resources
is beset with growing uncertainties in a fluid situation, in which
developing countries that produce resources are making nationalistic demands, advanced industrialized nations are moving to secure
resources at low cost and on a stable basis, and major international
interests are trying to strengthen their unity in the wake of the
threatened collapse of their oligopolistic organization.

Japan, of course, is most susceptible to these uncertainties
in the world's supply of resources. If this country continues to

rely heavily on the simple importation formula whereby mineral deposits mined by foreign enterprises are purchased as in ordinary transactions, it will become extremely difficult to secure a stable supply.

Japanese industry, moreover, is expected to be placed in a more difficult position than before in the procurement of production factors such as land, labor, and capital, as well as in the control of pollution. If resources are difficult to obtain, Japan's industrial development may receive a fatal blow.

International Aspects

The international aspects of Japanese natural resources needs impinge on

United States public and private interest. The White Paper observes:

Distribution of mineral resources on the international market tends to cause far more friction and provides far less flexibility than that of most other international commodities. Some of the major reasons for this are as follows:

i. Production of mineral resources is primarily not amenable to flexible operation. Since these are natural resources, their reserves are limited by the extent to which these underground resources can be mined with efficiency. In order to meet the growing demands new reserves must be discovered. However, considerable risks are still involved in the discovery of new deposits, despite the progress made in technology. A preparatory period of several years is normally necessary to find a new reserve. Moreover, natural resources, unlike other products, show only small changes in demand in relation to price fluctuations. This tends to make the producer very cautious in making a demand estimate. For this reason, the producer finds it difficult to vary output flexibly with changes in demand. Hence the perennial uncertainty over the possible shortage of raw materials.

ii. Production is concentrated in certain regions or countries, because reserves of natural resources are geographically localized. In addition, the tendency of concentrated production in certain parts of the world is enhanced by the fact that deposits of minerals liable to development technologically as well as economically are given priority in mine projects. Therefore, a change in the supply situation in a large producing area, combined with the rigidity inherent in resources production, tends to disturb their distribution.

iii. Most of the important mineral resources are under the control of major international interests.

(1) Major international firms normally hold control over key
mineral resources around the world because (a) mineral deposits
are concentrated in certain regions, (b) heavy risks are involved
in mining and development activities, (c) profit from mine development is inordinately large depending on the quality of the mine
involved, and (d) technology or commercializing the products can
be monopolized depending on the type of mineral mined.

(2) For these reasons, international distribution of mineral resources is virtually controlled by world enterprises. For example,
the seven so-called "major" or international oil interests control the supply of petroleum. The supply of copper is in the
hands of 10 international firms, while that of aluminum is governed by six major interests. These world enterprises launched
development projects abroad as early as 1900 and, taking advantage
of their overwhelming market share, formed cartels to regulate
prices of key mineral resources. Through these price cartels
major international interests were assured of their development
and established their worldwide distribution systems on an advantageous footing.

iv. Another important development is the rising nationalistic
sentiment in developing countries where mineral resources are produced or deposited. Since these raw materials represent a crucial
source of revenue for these countries, they tend to regard their
natural resources as an important national asset. Moves to sell
these assets as dearly as they can have been observed of late in
joint actions taken by members of the Conseil Intergouvernemental
des Pays Exportateurs du Cuivre (CIPEC) and the Organization of
Petroleum Exporting Countries (OPEC). Nationalism in these producing countries has also manifested itself in the expropriation
of foreign industries and mining concessions. These moves are
beginning to exercise a major influence on the distribution market
for mineral resources.

v. The factors mentioned above are intertwined, creating major
problems in the pricing of mineral resources on the distribution
market. Unless the world's demand for these raw materials is
substantially reduced in the future, consuming nations may continue to be plagued by uncertainties over the supply of these
essential resources.

The White Paper cites Japanese dependence for 10 major resources now at

90% and expected to grow to 93% in 1975. It goes on to note the overall

situation:

The world's resources supply structure is also expected to be confronted with difficulties in the 1970's. Since world demand will expand far beyond the level for the 1960's, uncertainties over the stable supply of raw materials will become more pronounced. Since the turn of this century the world's population has reached 3,500 million, and it is increasing at the rate of 60 million every year. Of course people around the world desire to enjoy more fruits of material civilization through elevation of their living standard. Reflecting this universal desire, per capita consumption of key mineral resources has been steadily rising. The population explosion and the rapid rise in per capita consumption have produced a multiplying effect on the world's demand for resources. For example, cumulative consumption of iron ores for the 1970's is expected to increase by about 50 percent over that for the 1960's. The estimated rate of increase for nonferrous metals is about 80 percent and that for petroleum about 120 percent.

Worldwide efforts to develop these and other essential underground resources will be required in this decade on a scale far greater than in the 1960's if the global supply of these raw materials is to be secured smoothly relative to the huge demand that is anticipated in the 1970's. In order to maintain the reserves-to-production ratio at the end of the 1970's on a par with that at the end of the 1960's, new mineral deposits twice as much as reserves discovered in the 1960's must be found. In the 1960's large oil fields were discovered in the Middle East but there are no assurances that such oil reserves will also be found in the 1970's.

Great risks involved in prospecting, enormous capital needed, and increasing physical, economic and political difficulties encountered in development projects do not allow of optimism concerning the supply of resources in the years ahead.

Circumstances surrounding the supply of resources are also subject to change in this decade, thus introducing additional elements of instability into the world's supply situation.

Such changes in the supply situation in Japan and in the world will make it increasingly difficult to obtain essential raw materials at reasonably low prices. In the days that Japan's imports of resources held a very small share in the world's trade in this field, raw materials were obtained from the existing world distribution market almost "by chance," i.e., without making elaborate predetermined efforts.

However, now that Japan is the largest importer of raw materials in the world and is expected to carry an increasing weight in the international distribution market, it can no longer

continue to depend as heavily on foreign suppliers as it has. If
this situation of excessive dependence or the lack of voluntary
efforts to develop resources on its own is not rectified, it will
become more and more unlikely for Japan to be able to secure raw
materials at low prices on a steady basis at a time when the
world's supply and demand relations are constricted by growing
uncertainties.

The report notes the rising nationalism in producing countries:

In recent years a tide of nationalism has swelled in developing
countries that produce natural resources constitute an important
source of revenue in developing countries/sic/, they demand a greater
profit share in the development and production of resources.
More specifically, they have come to request their capital participation in various projects and even decide to nationalize related enterprises in their territories. In other cases, they now
demand local processing and improvement of their infrastructure.

Such nationalism emerged in Mexico in the 1910's. Nationalistic demands were made at that time on a country-by-country basis.
The present tendency, however, is toward strengthening the unity
of developing countries that produce natural resources. The
rationale behind these moves originates from the United Nations
resolution concerning permanent sovereignty over natural resources
and from discussions that followed this resolution. This resolution found its practical meaning in the Organization of Petroleum
Exporting Countries (OPEC) established in 1960 and the Conseil
Intergouvernemental des Pays Exportateurs de Cuivre (CIPEC) organized in 1968.

These broad steps taken in the 1960's will be followed in the
years ahead by (a) more specific demands for capital participation
in processing, distribution, and marketing activities and (b) more
direct involvement in management decisions leading to production
restrictions and mandatory reinvestment.

Mineral resources are an outstanding international commodity.
From this point of view it is necessary for developing countries
and business interests of advanced nations to reach agreement on
the development and production, distribution and marketing of
these essential raw materials and fuels.

The growing desire on the part of advanced industrial nations
to develop mineral resources with their own capital and technology
and nationalistic moves among developing countries to promote

their own interests are expected to render extremely fluid the
international situation on resources. Principal among the
expected developments are the following:

i. A new age of resources development will come. As described
before, development projects have been undertaken mostly by a
handful of major international interests since these giant enterprises made their debut and expanded their operations abroad
around 1900. In the future, however, opportunities for participation of new entries in the development of resources will increase for these reasons: (a) the need to promote development
projects designed to ensure a rapid rise in demand and a stabilized supply in the 1970's, (b) the growing desire to develop
overseas resources on the part of advanced industrialized nations,
and (c) mounting nationalism in developing countries.

ii. These developments will also affect the supply and marketing
of resources and jolt the existing system.

iii. The need for mutual understanding between developed and
developing countries will be stressed more strongly than ever so
as to ensure sound distribution of the world's resources in the
1970's. To this end there will arise the need to establish order
in the development and distribution of resources.

     Duties and actions on the part of Japan, as a major consuming
nation and as the largest importer of natural resources, will
assume special importance in the changing international situation
of the 1970's.

Japan's difficulties in overseas development of resources are cited:

General problems:

i. Development projects involve great risks, and require heavy
initial investment and a long gestation period.

ii. Japan has little experience in overseas development. Most
of the present concessions are in the hands of major international
interests. Moreover, information and data needed to determine
the efficiency of development are lacking.

iii. Japan also lacks qualified manpower, and is also lagging
behind other advanced nations in many fields of technology.

Financial problems:

The most important problem is that Japan will need a colossal
sum of risk money in order to push overseas development projects,
considering that these ventures involve risks and require a long
preparatory period.

A rough estimate puts the financial requirement for the next five
years at about $5,000 million and that for the next 10 years at
$15,000 million.

On the basis of this estimate, the balance of direct investment
in resources development projects at the end of 1975 is expected
to represent 48 per cent of the nation's total balance of direct
investments which is estimated at $12,500 million. The balance
in 1980 would make up 59 percent of the estimated total balance
of $27,000 million. In other words, resources development is
expected to occupy the central position in Japan's overseas investment activities.

It is indeed no exaggeration to say that the largest problem in
overseas development of mineral resources is that of raising such
an enormous amount of funds.

The relationship between this need for capital and the probable

accumulation of huge surplus funds by oil-producing nations poses

interesting possibilities and problems.

The White Paper also discusses relations with developing countries in

terms most significant to the United States Government.

Japanese operations overseas must be guided by a policy of international cooperation so that Japan may contribute to establishing
order in the world's resources economy. The Government, for its
part, should take positive measures from a new point of view.

(a) Promotion of resources diplomacy:
Since resources problems, particularly those in developing countries,
transcend the boundary of private enterprise and pose questions
of grave interest to their states and government, government-level
diplomacy aimed at building friendly relations with developing
countries should be promoted. Also cooperation with enterprises
of advanced nations is important. The Government should strengthen
its guidance in order to promote such cooperation.

(b) Promotion of economic cooperation:
Economic aid designed to develop natural resources in a developing country or improve its related industries and infrastructure
can be very effective.

References

Abelson, P. H., 1972. "Environmental Quality," _Science_, v. 177, n. 4050

Benedict, R., 1967. _The Chrysanthemum and the Sword: Patterns of Japanese Culture_. (London: Meridian), 324 pp.

The Boston Consulting Group, Inc., 1972. _A perspective on Japan_. (Boston: The Boston Consulting Group, Inc.)

Dennison, C. S., 1972. "Summaries of Interviews," Exhibit 5 to _Japanese Environmental Situation and Resources Policy_, report prepared for Study Committee on Environmental Aspects of a National Materials Policy (unpublished), 133 pp. Interviews with M. Yasutake, Tokyo Electric Power Co.; K. Uchida, K. Nukazawa, and R. Wada, Keidanren Committee for Environmental Enhancement; Iwata, Economic Planning Agency; K. Arai and S. Nakao, Oil and Coal Bureau, Ministry of International Trade and Industry (MITI); H. Saegusa, Industrial Location Policy Division, MITI; T. Arai, The Dowa Mining Co., Ltd.; G. Mori, The Japan Mining Industry Association; T. Oyaki and Y. Sugiyama, Ministry of Foreign Affairs; S. Musakawa, T. Takeshita, and C. Okagawa, Institute for Policy Sciences; K. Matsuda, Japex Offshore Limited; H. Sato and S. Kudo, Japan Exploration Co., Ltd.; and H. Harada, Japan Resources Technical Institute.

Drucker, P. F., 1968. _The Age of Discontinuity_. (New York: Harper Row)

Environment Agency, Air Quality Bureau, 1972. _Air Pollution and Countermeasures in Japan_. (Tokyo: Environment Agency), 107 pp.

Forbes, 1972. "Volume not Profits," May 1, p. 26.

Halloran, R., 1969. _Japan; Images and Realities_. (New York: Alfred A. Knopf, Inc.)

Indonesian Observer, 1972. "Independent Body Needed for Environmental Management Says Wahjudi," Sept. 18, p. 32.

Institute for Policy Sciences, 1972. _An approach to the Ecosystem for Finding Ways to Harmonize Industrial Activities to the Nature_. (Tokyo: Institute for Policy Sciences), 16 pp.

International Development Center of Japan, 1971. _International Development Center of Japan_

Japan Mining Industry Association, 1971. _Copper, Lead, Zinc, Nickel_. (Tokyo: Japan Mining Industry Association)

Japan Resources Technical Institute (JRTI), 1972. Epitome of Japan
    Resources Technical Institute. (Tokyo: JRTI), 7 pp.

Japanese Office of the Special Representative for Trade Negotiations,
    1972. Japanese Government Procurement Policy. (Washington, D.C.:
    Japanese Office of the Special Representative for Trade Negotiations)

Kahn, H., 1970. The Emerging Japanese Superstate: Challenge and Res-
    ponse. (Englewood Cliffs: Prentice Hall)

Keidanren, 1972. Japanese Industry's Basic Attitude to Environmental
    Problem. (Tokyo: Keidanren), 9 pp.

Ministry of International Trade and Industry (MITI), Enterprise Bureau,
    1970. Laws and Regulations for Water Quality and Pollution Control.
    (Tokyo: MITI), 7 pp.

Musakawa, S., 1972. Systems Approach in Japan on Environmental Conserva-
    tion. (Tokyo: Institute for Policy Sciences), 14 pp.

Nagano, S., 1972. "Closer Japan-Chile Ties Seen in New Pacific Group,"
    The Japan Times, Sept. 18

The New York Times, 1972. "Japan and Soviet Agree on Joint Oil-Gas
    Project," Nov. 24

The New York Times, 1972. "Japan Helping Oil Industry Invest in Mideast
    Venture," Dec. 26

The New York Times, 1973. "Problems of Recycling," Feb. 2

Reischauer, E. O., 1964. Japan, Past and Present. (New York: Alfred
    A. Knopf, Inc.)

Sulzberger, C. L., 1972. "After the Ice Age in Japan," The New York
    Times, March 12, p. 11

K. Takashi, 1969. The Rise and Development of Japan's Modern Economy.
    (Tokyo: JiJi Press, Ltd.)

Tanaka, K., 1972. The Path for a New Japan: Summary of the Remodelling
    of the Japanese Islands (Tokyo: Background Information No. 71,
    Ministry of International Trade and Industry (MITI) Information
    Office), 11 pp.

Time, 1972. "Spare Clarity," Oct. 30

Touche Ross International, 1971. *Japan*. (New York: Touche Ross and Co.)

U.S. Department of Commerce, 1972. *Japan--The Government-Business Relationship*. (Washington, D.C.: U.S. Department of Commerce)

World Economic Information Services, 1972. *Economic Information File*. (Tokyo: World Economic Information Services)

World Environment Newsletter, 1972. "Japan Curbs Its Environment Agency," Oct. 24

Aesthetics, and coal mining, 111

Agard, J., 90

Agency for International Development (AID), 194, 195

Air, pollution of, 14; and coal mining, 111; by wood producing industries, 148

Air Pollution Control Administration, 42

Alaska, 133, 134; forests of, 157; and oil pipeline, 181

Alcohol, 72

Aluminum plants, 78

Aluminum smelters, 77

Anderson, H. W., 143

Animals, effects of fluorides on, 77-79

Arsenic, 74

Arctic Basin, protection of, 32

Asbestos, 74

Atmosphere, 197. See also Air

Backhoeing, 90

Balance of payments, 31, 171-172

BAT. See Best Available Technology

Benedict, H. M., 77

Benefit-cost analysis, 40, 55, 56

Benzo(a)pyrene, 74

Best Available Technology (BAT), 83; recommendations on, 86-87

Bidding systems, 55

Biosphere, 39

Black lung disease, 75, 76, 114

Bormann, F. H., 145

Boundary and Water Commission, United States and Mexico, 32

Brazil, 177

Bronchitis, 75

Brown, G. W., 144

Building materials, recovery of, 183

Bulk transport, 180. See also Transportation

Bullet trains, 208

Bureau of Indian Affairs, 115

Bureau of Land Management, 115

Bureau of Mines, 94; systematic projections of, 169

Cadmium, 170

Cadmium poisoning, 210

Canada, arctic and, 201

Cancer, lung, 74. See also Black lung disease

Carcinogens, 62, 72, 74

Catastrophe, environmental, 36

Change, technological, 58-61

Chemicals, on-the-farm use of, 36; recovery of, 183; synthetic organic, 169

Chromates, 74

Chromium, 170

Chromosomes, aberrations of, 77

Cities, environment of, 6

Clean Air Act, 3, 28, 191

Clearcutting, 142, 144, 145, 158. See also Logging

Coal, mining of, 111-113; production of, 107-110; reserves of, 107-110; source of, 7; sulfur content of, 109; utilization of, 115-117

Coal dust, 114

Coal Mine Health and Safety Act, 28, 75, 114, 115

Coastal zones, 10, 13

Coast Guard, United States, 119

Committee on the Challenges of Modern Society (CCMS), 189

Committee for International Environmental Programs (IEPC), 229

Commuting, 60

Conference on the Law of the Seas, 197, 198, 201

Conifers, and hydrogen fluoride, 78; and SO_2, 79. See also Forests

Conservation, land and water,
17-18

Consumption, 7, 63; per capita
energy, 105

Container industry, 21

Containers, problem of, 58

Continental shelf, 197; oil
drilling in, 118

Contracts, private, 45

Convention on the Continental
shelf, 198

Convention on the Prevention
of Marine Pollution by Dumping
of Wastes and Other Matter,
198

Copper, 57

Corporations, multinational, 33,
179-180, 196

Costs, of environmental damage,
5; environmental policy, 40;
of health, 75-76

Council on Environmental Quality,
43, 172, 191

Conseil Intergouvernemental des
Pays Exportateurs du Cuivre
(CIPEC), 219, 221

Countryside, environment of, 6

Critical boundaries, concept of,
139

Crude oil, 182. See also Oil

Curry, R. R., 145

Damage functions, 61

d'Arge, R. C., 171, 172

DDT, 177

Demand, for mineral resources,
68

Department of Health, Education
and Welfare, 79

Design standards, 20, 48

Desulfurization, 213

Developing countries, 177-179,
193, 202; nationalism in, 221

Discharges, liquid, recovery
from, 183

Disease, 71; chronic, 80; heart,
75; respiratory, 75; and total
body burden, 72-73

Disease control, 12

Diseconomy, external, 40, 41

Disposal, 10, 39; of spent shale,
122

Dochinger, L. S., 80

Ecological studies, 93

Economic Commission for Europe
(ECE), 189

Economic Planning Agency,
Japanese, 210, 211

Economy, and coal mining, 111;
external, 40; international,
163, 166

Education, and environmental
problems, 30

Effluents, 146; charges on, 50;
standards for, 16, 47

Egypt, 177

Electrical industry, 111

Electricity, generation of, 105

Emissions, charges on, 49;
industrial, 35; recovery from,
183; regulations on, 181;
standards for, 16, 47, 49, 59;
taxes on, 15-16, 49; transport
of, 35

Emphysema, 75

Endangered species, 195

Endangered Species Conservation
Act, 28

Energy, economical use of, 60;
estimated resources, 101; geothermal, 124-125; new forms of,
36; nonfossil sources of, 124-
128; solar, 126

Entrepreneurs Bearing of the Cost
of Public Pollution Control,
Japanese, 209

Environment, control of, 126
(See also Environmental
control); effect of materials
flow on, 13; and energy
problem, 101; and federal
legislation, 27; forest, 140
(see also Forests); and human
health, 71; realistic standards for, 105-107; threat to,
3

Environment Agency, Japanese,
210, 211

Environmental control (EC), 166;

costs of, 168, 169; domestic norms, 169; international, 177; and international trade, 174, 175; national, 29; national implications of, 176; and price shifts, 171; standards of, 175
Environmental impact statements, 34, 117, 138; of Department of the Interior, 119; requirements for, 194
Environmental policy, development of, 42. See also Policy
Environmental Protection Agency (EPA), 48, 191
Environmental quality, specifications for, 137-141
Environmental Studies Board, 229
EPA. See Environmental Protection Agency
Epidemiological studies, 35, 72, 74
Erosion, 71; in forests, 142
Eutrophication, 146
Exploration, 26, 107; gas, 118; oil, 118; techniques of, 90
Export-Import Bank, 195
Extraction, 26, 31, 39, 107; mineral resource, 68; of nuclear fuels, 124
Federal Aid Highway Act, 28
Federal Insecticide, Fungicide and Rodenticide Act, 28
Federal Power Act, 28
Federal Power Commission (FPC), 116, 123, 124
Federal Water Pollution Control Act, 28; Amendments of 1972, 3, 149, 191
Federal Water Pollution Control Administration, 42
Fertilizers, 17, 146; recovery of, 183
Fibers, 10
Finding rate, for oil, 118
Fish and Wildlife Coordination Act, 28
Fisheries, sport, 142
"Flag of convenience," 199
Flawn, P. T., 89

Fluorides, 73, 76; effects of, 77-79
Food and Agriculture Organization (FAO), 166
Food, Drug and Cosmetic Act, 28
Food chain, 13, 71, 139, 184
Forestry, 35; Canadian, 153-154; international, 151; legislative restrictions on, 137; management in, 155; recommendations for, 157
Forests, 19, 133; damage to, 141, 142-144; exploitation of, 151; fires in, 140; harvesting of, 152, 154, 157; land management of, 138; products of, 132; regeneration of, 145; state, 137; timber production, 136; tropical, 13
Fredriksen, R. L., 145
Fuels, 101; materials policy for, 23; nuclear, 123; on-the-farm use of, 36
Furs, 167
Fusion, 125
Gas, liquid natural (LNG), 6, 168, 182, 213; synthetic natural (SNG), 116
Gasification, of coal, 117
GATT, 169
Ghetto, urban, 7
Gobs, 114
Gordon, C. C., 78, 79, 80
Gorham, E., 79
Grasses, and hydrogen fluoride, 78
Growth, 3; economic, 4, 61-63
Harvesting, 26; controlled, 168
Haydu, E. P., 146
Health, 76; costs of, 75-76; economic, 71; environmental, 71; environmental effects on, 12-13; human, 71; metal mine, 75; of miners, 114-115; recommendations on, 80-81; standards for, 25; surveil-

lance and, 74
Health studies, 35
Hearing, loss of, 74
Heart attacks, 75
Heat, exposure to, 73
Heat balance, 62
Heavy metals, 15
Hedgecock, G. G., 79
Helvey, J. D., 142
Herbicides, 146, 158
Hewlett, J. D., 142
Hickle report, 125
Hot water deposits, 124
Hubbert, M. King, 118
Hydrocarbons, 72
Hydroelectric sources, 124
Hydrogen fluoride (HF), 77
Hydropower, 13
IEPC. See Committee for International Environmental Programs
IMCO, 199
Imports, 4, 32; restrictions on, 195
Income, inequal distribution of, 7
Incineration, disposal by, 147
India, 177
Indonesia, 177
Industry, export, 170; geothermal, 125; Japanese, 218; minerals, 70, 95; mining, 73; nuclear fuel, 123; and pollution, 48; and recovery efficiency, 82; relocation of, 208; wood producing, atmospheric emissions from, 148
Information, on environmental degration, 178; technical, 153
Information bank, 90, 91
Insecticides, 146
Instruments, policy of, 43-45
Intergovernmental Maritime Consultative Organization (IMCO), 198
International Biological Program, V General Assembly of, 153
International Boundary and Water Commission, 192

Internationalization, 29
International Joint Commission (IJC), 192
International Seabed Authority, 199
International Standards Organization (ISO), 166
Investment, government, 51
Itai-Itai disease, 210
Japan, 33; energy strategy of, 212; environmental problems of, 209; industrial pollution in, 210; international cooperation of, 223; natural resources of, 217; postwar development of, 207; raw materials for, 214; resources of, 222; sulfur reduction in, 212; and United States, 215
Kneese, A. V., 171, 172
Krammes, J. S., 143
Krutilla, J. V., 92
Krygier, J. T., 144
Land, and coal mining, 111; nonindustrial private, 156-157; and oil shale processing, 121; surface, restoration of, 115; surface mined, rehabilitation of, 113; use of, 5
Land and Water Conservation Fund Act, 28
Landslides, 142, 143
Land tenure, 88
Land use, planning, 88
Lantz, R. L., 144
Lave, L. B., 123
Law, environmental, 30; international, 31, 204
Law of the Seas Conference, 197, 198, 201
Lead, 72, 170
Lead poisoning, 76
Leaffall, 145
Leasing, energy resource, 127; shale, 123
Leasing Act of 1920, 88
Leasing policies, federal, 115
Legislation, clean water, 47;

domestic, 189; and explo-
ration, 91; federal, 26, 27,
28; and materials policy, 25
Liquefaction, of coal, 117
Liquified natural gas (LNG), 6,
168, 182, 213
Litigation, 45, 46
Litter, 14, 17; synthetic
organic, 169
Littoral state, 197, 200
Living patterns, 36
Living standard, Japanese, 217
LNG. See Liquid natural gas
Location element, 52
Logging, 143, 144, 145. See
also Forestry
Lull, H. W., 142
Lumber, source of, 7
Lumbering, 17. See also
Forestry
Manpower, 86; Japanese, 222
Marine Protection, Research, and
Sanctuaries Act, 28, 199
Marine safety, 180
Market system, 39. See also
Economy
Materials flows, environmental
effects of, 10-14
Materials policy, 2-10, 39; basic,
184; and environmental policy,
136-137; international aspects
of, 189; national, 5
Materials supplies, 180
Materials utilization cycle, 69
Megalopoles, 6
Mercury, 170
Mercury poisoning, 210
Metals, 10, 170
Metal scrap, export of, 183
Mexico, nationalistic demands
of, 221
Minamata disease, 210
Mine drainage, 17
Mine Health and Safety Acts, 81
Mineral Leasing Act, 28
Mineral resources, international
control of, 218, 221
Mines, "re-packing" of, 114;
underground, safety of, 127

Mine wastes, disposition of,
113-114
Mining, 17, 55; and BAT, 83; of
coal, 111-113; environmental
impacts of, 111; health and
safety of, 114-115; ocean, 70;
of oil shale, 121; reclama-
tion, 93; right to, 88; siting
of, 91; surface, 122; under-
ground, 122
Mining Law of 1972, 88
Mining and Minerals Policy Act,
28
Ministry of Agriculture,
Japanese, 211
Ministry of Construction,
Japanese, 211
Ministry of Health and Welfare,
Japanese, 211
Ministry of International Trade
and Industry (MITI), 210, 211
MITI. See International Trade
and Industry Ministry,
Japanese
Mitsui Mining and Smelting
Company, 209
Mohammed, A. H., 77
Monitoring, 34-37, 74; of
environment, 6; of forests,
138; global, 35; of oil shale
mining, 123; technology of,
139
Multinational corporations, 33,
179-180, 196
Multiple-Use Sustained Yield
Act, 28
National Academy of Engineering,
228
National Academy of Sciences,
228
National adjustment assistance
program, 174
National Coal Association, 116
National Commission on Materials
Policy, 102, 132
National energy policy, 26
National Environmental Policy
Act, 3, 5, 28, 52, 191, 194,
195, 196; and continental

shelf, 198; impact statement procedure of, 44
National Oceanographic and Atmospheric Administration, 191
National Park Service Act, 28
National Petroleum Council, 103, 123, 125
National Research Council, 228
Nationwide Forest Survey, 134
NATO, 29
Natural gas, potential resources of, 23; production of, 117-119; reserves of, 117-119
Nickel, 74, 170
Nixon, Richard M., 6
Noise, effects of, 74
Noise Pollution Control Act, 3, 191
NTA (nitrilotriacetate), toxicity of, 85
Nuclear power, Japanese, 213
Observation, systematic, 63. See also Monitoring
Obsolesence, 9, 27, 36
Occupational Safety and Health Act, 28
Oceans, 10, 13, 197-201; dumping in, 199; pollution of, 164; protection of, 32
Office of Emergency Preparedness, 102
Offshore drilling, 119
Oil, production of, 117-119; reserves of, 117-119; transportation of, 119
Oil Pollution Act, 28
Oil shale, 119-122
Oil spill, 14, 62, 119, 180
Only One Earth concept, 2
Organization for Economic Cooperation and Development (OECD), 189, 216
Organization of Petroleum Exporting Countries (OPEC), 219, 221
Orphan areas, 93
Overseas Private Investors Corporation (OPIC), 195
Ozone, 76

Packaging, 9, 21
Pakistan, 177
Paley, William S., 37
Paper, export of, 183
Paper mills, 15
People, effect on environment, 8
Permits, 45, 48-49
Pesticides, 17, 146, 158
Petroleum, demand for, 103; potential resources of, 23; source of, 7; sulfur, 6
Petroleum products, 197
Pharmaceuticals, recovery of, 183
Phenols, 148
Philippines, 177
Phosphate plants, 77, 78
Phosphorus, 11
Photosynthesis, and SO_2, 80
Planning, land use, 5, 19-20, 88
Plantations, 151
Plant ecosystems, artificial rehabilitation of, 107
Plants, 76
Plastics, substitution of, 84
Plywood plants, solid waste from, 147
Pneumoconiosis, 114. See also Black lung disease
Policy, economic, 26; energy, 101; energy fuel, 126; energy resource, 127; environmental, 42; federal land use, 115; instruments of, 4, 45; international, 202-204; materials, 2-10; national energy, 26, 102, 103, 104, 105; no development, 106; pricing, 127; trade, 6
Policymakers, 44
Pollutants, 8; air, 76; international commission for, 185;
Pollutees, 41
"Polluter-pays" principle, 31, 165, 166, 171, 172, 178, 216; international, 193; in Japan, 209; of OECD, 189
Polluters, 41

Pollution, air costs of, 77; industrial air, 75; and international law, 204; location element of, 53; marine, 181; transnational, 165, 184

Pollution control, costs of, 172; equipment for, 170; intangible benefits of, 55-57; in minerals industry, 95

Pollution havens, 176

Population, increase of, 36; Japanese, 209; pressures of, 4; "survival", 73;

Polychlorinated biphenyls, 177

Ports, restrictions of, 181

Ports and Waterways Safety Act, 119

Power plants, siting of, 20

Precipitation, chemistry of, 79

President's Council on Environmental Quality, 70

President's Materials Policy Commission, of 1952, 37

Pricing systems, two-part, 170

Primary production, 57

Processing, 10, 31; and BAT, 83; mineral resource, 68; of nuclear fuels, 124; oil shale, 121

Processing plants, siting of, 92

Production, 39; increase of, 36

Progress, technological, 59

Prohibitions, 45

Property, and materials policy, 25

Quality Conservation Law, Japanese, 209

Questionnaires, 56

Radioactive materials, 74

Rags, export of, 183

Rainfall, acidification of, 11, 71

Reclamation, definition of, 93; governmental, 51; requirements of, 93

Recycled materials, international trade in, 183

Recycling, 4, 10, 21, 57-58, 83-85; economic feasibility of, 22; primary, 150; water, 149

Refining, of coal, 117

Refuse Act, 49

Reforestation, 141

Rehabilitation, 127

Reinhart, K. G., 142

Remedies, for preventing environmental damage, 14

Research, 34-37; and environmental problems, 30; needs for, 107

Reserves, protected, 89

Residuals, controls on, 95

Resources, allocation of, 34, 40; "common property," 41, 49; nonrenewable, preservation of, 31; renewable, 24; consumption of, 132-133

Resources Recovery Act, 115, 148, 191

Reynolds Metals Company, 214

Rice, R. M., 143

Roads, construction standards for, 143

Runoff, water, 141

Ruston, A. G., 79

Safety, 71; coal mine, 75; marine, 180; of miners, 114-115; standards for, 25

Salinization, of Colorado River, 192

Salmon, 144

Sawmills, solid waste from, 147

Science, and environmental problems, 30

Science and Technology Agency, Japanese, 211

Scientific Committee on Problems of the Environment (SCOPE), of International Council of Scientific Unions, 43

Seabed, 82, 197, 198, 199, 200

Seas, pollution of, 31. See also Oceans

Sedimentation, 71; in forests, 142

Sewage, recovery from, 183; shipboard, 181

Shale, oil, 117, 119-122

Silviculture, 140, 141

Singapore, 177
Siting, conflicts in, 91-92
Skins, 167
Smelters, copper-nickel, 79
Smoke and Soot Regulation
    Law, Japanese, 209, 210
Smoking, 72
SNG. See Gas
Socioeconomic status, 72
Soil, artificial rehabilitation
    of, 107; fertility of, 12; and
    SO$_2$, 79
Solar systems, 126
Solid waste, 85-86; disposal of,
    21-22; dumping of, 181; of
    timber processing, 147
Solid Waste Disposal Act, 28, 191
Solvents, 72
Standards, antipollution, 191;
    design, 20, 48; effluent, 45,
    47-48; emission, 47-48; environ-
    mental, 69, 164; of health, 72,
    73-74; for lead, 76; product,
    167; realistic, 105-107
State, timber inventory of, 135
Steam, 124
Stockholm conference. See United
    Nations
Streams, temperature of, 144
Stresses, environmental, 8-9
Stressors, environmental, 72
Strip mining, 52, 88. See also
    Mining
Subsidies, 114, 178; direct, 45,
    46
Substitution, 83-85
Sudbury district, Ontario, 79
Sulfation, atmospheric, 71
Sulfur, reduction of, 212; removal
    of, 112
Suflur dioxide (SO$_2$), 53; in coal,
    109; emissions of, 148; emission
    standards for, 112; industrial,
    72
Sulfur oxides, 76; effects of on
    vegetation, 79-80
Supplies, allocation of, 34
Surface mining, coal, 103; regu-

lation of, 113
Surveillance, and health, 74
Swanston, D. N., 143
Swedish Royal Ministry for
    Foreign Affairs, 79
Tanaka, Prime Minister, 208
Tankers, oil, 119
Tariffs, 166
Tax, 58, 164; emissions, 15-16,
    49; materials, 22; pollution,
    185; privilege, 58
Taylor Grazing Act, 28
Technology, 8; assessment of,
    81-82; change in, 58-61;
    dependence on, 36; energy,
    214; and environmental
    problems, 30; exploration, 91;
    monitoring of, 139; reclama-
    tion, 104
Technology Assessment Act, of
    1972, 5
Technosphere, 39
Teepee burners, 147
Thorium, 123-124
Thut, R. N., 146
Tides, 126
Timber, 132; growth-removal
    balance of, 134-135; harvest-
    ing of, 134, 157; and logging
    practices, 24; management of,
    136; processing of, 146-150,
    154. See also Forestry;
    Logging
TLV (threshold **limiting value),**
    73
Tokyo Gas Company, **213**
Total body burden, **and disease,**
    72-73
Trade, international, **184;**
    world, 176
Tradeable materials, **recovery**
    of, 182-184
Trade Expansion Act, **173**
Trade flows, 174-176
Traffic, 60
Trailings, uranium, 94
Trail Smelter arbitration, 191
Transportation, 31; bulk, 180;

costs of, 182; high-speed, 208;
  oil, 119
Treshow, M., 77
Tropical forests, 151, 152
Truman, Harry S., 197
Uncertainty, question of, 62
United Nations, 204; Committee
  on the Sea Bed, 199;
  Conference on Human Environ-
  ment, 2, 29, 30, 33, 151, 189,
  184, 190, 202-204; Environ-
  mental Secretariat of, 201
United States, coal reserves of,
  108; environmental policy of,
  31-33; and international
  organizations, 190; and Japan,
  215; multinational corporations
  of, 33
United States-Canada International
  Joint Commission, 32
University of Alaska, 125
Uranium, 123-124; enriched, 214
Use, 63; mineral resource, 68;
  land, 5
Vanadium, 74
Vannier, M., 90
Vegetation, 76; effects of
  fluorides on, 77-79; effects
  of sulfur oxides on, 79-80
Walter, I., 175
Waste disposal, 12
Waste heat, discharge of, 11
Wastes, conversion to, 148;
  dispersion of, 16-17; mine,
  22, 113-114; recovery from,
  183
Water, chemical quality of, 145-
  147; pollution of, 14; recy-
  cling of, 149
Water pollution, by wood pro-
  cessing industries, 148
Water Pollution Control Act, 46
Watershed, 133
Water system, and coal mining,
  111
White pines, chlorotic dwarf
  disease of, 80
Wild and Scenic Rivers Act, 28
Wilderness, 19, 133

Wilderness Act, 28, 191
Wildfires, 137
Wilkes, H. G., 146
Winds, 126
Wood, 132; export of, 151;
  waste, 147, 148. See also
  Timber
Wood processing industries,
  aqueous wastes from, 148
Wood producing industries,
  atmospheric emissions from,
  148
Work load, 73
World Bank, 196
World Health Organization, 76
Zablotney, J., 119
Zero growth, 62
Zoning, 92